UNCHAINED

Your AI Blueprint for Liberation

Mark Mueller

MM

Mark Mueller

An Imprint of Mueller Publishing

All rights reserved. Published by Mueller Publishing, Santa Rosa, CA 95407.

ISBN: 979-8-9938219-0-0 (eBook)
ISBN: 979-8-9938219-1-7 (Hardcover)
ISBN: 979-8-9938219-2-4 (Paperback)
ISBN: 979-8-9938219-3-1 (Audio)

Cover Design: Generated with AI and significantly edited/finalized by Mark Mueller.

For ALL who dream of being

"Unchained"

Table of Contents

Preface

The idea for this book didn't come to me in a classroom, a boardroom, or even from poring over financial reports. It hit me on a casual walk down a familiar trail near my home, during one of the most challenging periods of my life. For two years, I was out of work, a casualty of a tech startup layoff, pursuing every opportunity, only to be passed over time and again. As my unemployment benefits dwindled, the reality of my situation grew heavier. I was also tethered by an unwavering commitment to two little Jack Russell Terrier-Chihuahua mixes – my children, really, along with a partner who had to take on the burden of caring for us financially. My self-worth began to plummet, and I started to see a pattern of constant rejection and a growing sense of hopelessness. I was starting to question myself, my skills, and my future.

One of my fur babies, my sweet boy Bowie, had been diagnosed with a chronic illness. His kidney disease was the initial diagnosis, but it was his high blood pressure that caused a painful case of glaucoma. For a dog that could not speak, this was his silent hell. It demanded a relentless schedule of heart medication and the four daily eye drops his glaucoma required, a regimen that left me consumed. I gave my soul to this dog, and when his high blood pressure finally caused his eyes to necrotize, leaving him completely blind, his reliance on me became absolute. Most people would have put him down, but he was still eating, still wagging his tail, and still wanted to go for his walks. It was in this devotion—cooking him homemade food, administering his medication, and tending to his every need—that my own day-to-day existence was defined by a single purpose: honoring his life.

But even in that profound devotion, the system's relentless pressure found its way in. My life was reduced to the balance in my bank account. The crushing anxiety was relentless, and my identity, my hope, and my very sense of self were reduced to a simple equation of money in versus money out. My body began to shake, my mind clouded over, and I started experiencing heart palpitations—symptoms so severe I had to take anxiety medication just to function. I was even put on medication for atrial fibril-lation. It was in this complete state of mental and physical collapse that I had a chilling realization: At that stage, I was part of a system built for my calculated destruction that had direct control over my life, my freedom, and my future.

In that same suffocating period, my relationship was enduring its own storm. With my partner shouldering the immense financial burden, an unspoken stress and resentment festered between us. The shame within me grew not from a lack of effort, but from the feeling of profound failure. Even while I dedicated myself to caring for our dogs and relentlessly searched for work, I felt as if I was failing to re-integrate into the system's demands fast enough, leaving me trapped in a horrible cycle of rejection. The shame felt even deeper because, on top of this, the truth was, I've always found comfort in my own mess, and a childhood without the responsibility of chores had left me unprepared for the demands of a home. My own inner turmoil, exacerbated by being out of work, left me frozen and incapable of stepping up when my partner needed me most. Unable to articulate his frustrations, he occasionally succumbed to emotional outbursts. While I was interviewing constantly and facing rejection after rejection, the futility of my efforts seemed to disappear into a void. Even though he witnessed my exhaustive work, I could feel the silent, unspoken judgment from his end—a frustration not at my lack of trying, but at my lack of results. We were both consumed by our own private hells, unable to connect, as the relentless pressure of the system pushed us further and further apart. I still flinch at those raw memories, but what I know for certain is that our love, and the strength we found in it, was the only thing that allowed us to sur-vive that treacherous time.

In that moment on the trail, confronting the weight of my life in ruins, the thought struck me with the force of a physical blow, a gut-punch of empathy that was almost too much to bear: the terrifying and unfathom-able reality of being a parent, out of work, with empty stomachs and a sick child relying solely on a bank account balance for a roof, food, and medicine. It was a horrifying truth: the very survival of a family, a child's health, and their next meal were reduced to nothing more than a number on a bank account screen. This was no longer just about me; it was about the system itself. The weight of that thought, the visceral fear of being so utterly at the mercy of my bank account balance, was crushing. And it was the pressure of that balance that cracked open my understanding of the world, revealing the invisible chains of a system that acts as a parasite on our very existence—effectively, slaves to a machine that demands our labor while offering little true security or humanity.

This realization ignited a fierce anger within me. But alongside it, a powerful insight began to emerge. Having worked at several startups, two of which were pioneers in machine learning and artificial intelligence, my eyes had already been opened to the vast possibilities of AI. I knew, with absolute certainty, that we should not be afraid of this technological revolution, but rather embrace it. Yet, the status quo, the corporate machines, seemed intent on ensuring that the majority remained unprepared, perpetually at their mercy.

It was then that I connected the dots: my own past financial ignorance, a condition that plagues so many from a young age, and the recurring cycles of debt and painful instability that put a profound strain on my relationship, as well as the friendships I lost in those tumultuous years. It hurt deeply, because those relationships meant so much to me, even as a core group of friends stood by my side. All of this, I now saw, stemmed from a lifetime of systemic conditioning and a fundamental misunderstanding of the power of numbers—not just in a bank account, but in shaping one's entire life and relationships.

All of these experiences—the relentless rhythm of financial instability that defined my life, the stress that eroded my self-worth, the loss of friendships, and the agony of watching my partner work like a slave to keep us afloat—came together in a single, searing moment of clarity. **This is why I vowed: I would not let anyone else live through that hell if I could help them.** This book is that help. It is born from that anger, that fear, that love, and the profound realization that we have been robbed of financial security. This is not just an economic theft; it is an emotional and psychological violation. But a better life is possible, not because money buys happiness, but because it provides the comfort and freedom to avoid agonizing decisions, easing the burdens on yourself and those you love. It allows you to build a life where financial well-being supports, rather than sabotages, your most important connections. This book is your guide to breaking free from the old system, understanding the new one, and leveraging every tool to become the architect of your own destiny, reclaiming the life that was stolen from you. **This is your liberation.**

Part I:

The System of Servitude

Chapter 1:

The Great Divide: A History of Betrayal

The idea that some people are rich and others are poor is hardly new. Look back through history, and you'll find wealth disparity woven into the fabric of societies, from ancient empires to the gilded age of the American Industrial Revolution. Names like Rockefeller, Carnegie, Edison, and Vanderbilt still echo from a time when colossal fortunes were built, starkly contrasting with the grueling lives of the working masses. Back then, efforts were made to temper this divide; the push for the eight-hour workday, famously embraced by Henry Ford, was a testament to the idea that stability and a fair living wage for workers were crucial for a functioning society. The common wisdom suggested that free markets, guided by an "invisible hand," would naturally distribute wealth and opportunity for the good of all.

But while inequality has always been a companion to human civilization, what we've witnessed since the early 1980s is an entirely different beast. The gap between the ultra-wealthy and everyone else isn't just widening; it's a chasm, deepened and solidified by deliberate design. The true inflection point, the moment the scales truly tipped, can be traced back to the early part of the Reagan Administration. A subtle but seismic shift in corporate regulation, particularly around executive compensation, opened the floodgates. Specifically, changes in SEC rules and tax laws made it far more advantageous for corporations to reward their top executives with stock options.

The 1980s: The Foundation is Laid

The 1980s set the stage. For decades, the unspoken rule was that a company had a broader responsibility beyond just profit—to its employees, its customers, and the community it served. But that social contract was slowly and deliberately poisoned by a parasitic plot, a cynical betrayal of humanity itself, hatched by a syndicate of Wall Street's most influential and ruthless architects. They would use the very technologies meant to advance humanity—the explosive power of personal computers, software, and early digital networks—to manipulate the very rules of the economy and subdue the populace into a new form of financial slavery.

This "plot" was a push for deregulation and a direct removal of the safety rules that had been put in place after the 1929 stock market crash to protect people's investments and livelihoods, effectively turning the stock market into a vast casino floor. The goal was to remove what were called "burdensome barriers," like rules that kept regular people's checking accounts separate from Wall Street's high-risk gambling and protected companies that wanted to remain independent from being taken over. This new philosophy, masquerading as a "free market," was a brutal, engineered fraud. It didn't "unleash" the market for everyone; it simply ripped the leash off corporations and let them run wild over the very safeguards that protected the public. This new corporate gospel demanded that CEO pay be directly tied to the company's stock performance. The blueprint for this fraud—a corporate mindset specifically engineered to justify limitless wealth for top executives—was now in place. While this was perfectly legal and some companies were already embracing it, it was far from the universal, default strategy it would become. All it needed was a legal excuse to turn that cultural trend into an unstoppable force.

The 1993 Policy: The Floodgates Open

The blueprint for financial slavery that was forged in the 80s became a brutal reality in 1993. The Omnibus Budget Reconciliation Act, championed by the Clinton administration, was sold as a progressive win—a law that would curb the public outrage over excessive CEO pay. This was a lie. The bill, passed by Democrats who were in control of the House, Senate, and presidency, was framed as a tax on the rich. It capped the deductibility of CEO salaries at $1 million, a move meant to look like a crackdown. But it contained a devastating loophole, a Trojan horse written by those same architects: any pay based on "performance" was exempt from the cap.

The effect was immediate and catastrophic. This policy didn't curb CEO pay; it detonated it. It turned the new corporate gospel of "shareholder value" into a legal and financial mandate. Now, every board had a powerful, tax-based incentive to stop paying CEOs in cash and start paying them in stock options. This was the moment the floodgates opened, and the unstoppable force went from a blueprint to a full-blown, undeniable reality.

The Ultimate Betrayal

The political theater surrounding this bill makes the betrayal even more profound. The final vote in the Senate was split 50-50, with Vice President Al Gore casting the tie-breaking vote to pass the law. This wasn't a righteous stand for the people; it was a political spectacle woven to create the illusion of bitter, party opposition. The Republican resistance was driven by their ideological stand against tax increases and government spending, not by any genuine awareness of the parasitic plot hidden within the text.

So here is the horrible truth: The Democrats, in their zeal to pass a law that appeared to tax the rich, unwittingly handed Wall Street the keys to a machine that would create financial slavery. The Republicans, in their equal zeal to block the Democrats, remained silent on the very provision that would devastate the people they were supposed to represent. Both sides, for their own reasons, were either complicit or utterly blind to the real war being waged. The public was sold a lie, and the political system, in all its tribalism and noise, was entirely unable or unwilling to stop it. The betrayal was a bipartisan act, with both parties ultimately serving as enablers for the architect's plot. And in the decades since, neither party has ever made a serious effort to repeal this mandate—proof that the true corruption runs so deep that the devil pulling the strings is not a person, but greed itself, a gluttonous beast that feeds on our collective well-being.

Chapter 2:

The Cold, Hard Proof: Quantifying the Divide

If the deliberate design of our economic system feels like a betrayal, the raw numbers are its cold, hard proof. These aren't just abstract figures; they represent millions of lives impacted, opportunities lost, and dreams deferred. They illustrate, with brutal clarity, the vast chasm that separates the ultra-wealthy from everyone else. Let's look at the evidence:

- **The Unimaginable Concentration of Wealth:** Consider this stark reality: The wealthiest 1% of households in the United States currently hold roughly one-third of the nation's total wealth. This isn't just a large slice of the pie; it's practically swallowing the whole thing. To put it another way, the roughly 1.3 million households at the very top control as much wealth as the entire American middle class and working class combined. Meanwhile, the bottom 50% of the population—over 165 million people—collectively own a meager 2.5% of the nation's total wealth. So while most of us are fighting for crumbs, the 1% are hoarding the entire bakery.
- **The Exploding CEO-to-Worker Pay Gap:** Remember our discussion about the shift in corporate priorities and the engineered greed of the 1980s? The CEO-to-worker pay ratio paints a vivid picture of its consequence. In 1965, the average CEO of a major U.S. corporation earned about 20 times what their typical worker did. It was still a gap, but one that hinted at a shared enterprise. Fast forward to today, and that ratio has skyrocketed. In 2023, CEOs at the top U.S. firms were paid, on average, 290 times as much as their typical worker. Some companies demonstrate even more egregious disparities; at certain major retailers, it would take a median employee THOUSANDS of years to earn what their CEO makes in just one.

- **The Burden of Debt:** While the ultra-wealthy accumulate assets at an unprecedented pace, a significant portion of the population is trapped under a rising tide of debt. Total U.S. household debt has reached over $18 trillion, with significant portions in mortgages, credit cards, and student loans. For many, debt isn't a strategic financial tool; it's a constant burden, a monthly struggle to stay afloat. This debt limits choices, stifles mobility, and perpetuates the cycle of financial insecurity, making it nearly impossible to build the kind of generational wealth that those at the top inherit.

- **The Intersectional Layers:** While these numbers reflect a national trend, the divide is even more pronounced when we look at specific demographics. For example, in 2022, the median White household held over $284,000 in wealth. Compare that to the median Black household, which held just over $44,000, or Hispanic households at around $62,000. These stark differences underscore that the system's design doesn't just benefit the rich; it systematically disadvantages certain groups, often along racial and ethnic lines, exacerbating pre-existing inequalities.

These aren't just statistics; they are symptoms. They are the measurable outcomes of the deliberate choices made by those in power, choices that have transformed our economic landscape into a rigged game. Staring at these figures, we reveal the system is not broken—it's meticulously engineered to extract our very soul.

Consider a high school student skipping lunch because their parents simply can't afford it. Or families losing access to critical healthcare because vital safety nets are stripped away, all while the financial elite protect and grow their obscene fortunes. It's not just an abstract policy; it's the very lifeblood being squeezed out of ordinary people, their protections dismantled piece by piece, their foundations eroded.

This isn't just indifference. It speaks to a chilling lack of empathy, a profound disconnect from human suffering that, to the average person, can only be described as a form of sociopathic control. The decision-makers, insulated by layers of wealth and privilege, implement policies that cause immense hardship, then stand back and watch the poor scrape by with scraps—sometimes, it seems, almost gleefully. The quantifiable data screams these people simply don't care. Their priorities lie solely with maximizing their own gain, even if it means sacrificing the well-being, health, and dignity of millions. The terrifying reality is that this pervasive, deeply uncaring behavior is precisely what holds the reins of our collective future.

If this is the reality, if the very system is designed to crush us, how do we, as individuals and as a generation, navigate it and truly break free?

Chapter 3:

The Human Cost: A Sickness of Society

The chasm between the ultra-wealthy and everyone else isn't just about staggering numbers on a spreadsheet or a debate in the halls of power. Its true brutality is felt in the everyday lives of millions, systematically eroding what was once considered the bedrock of a stable society: the quality of life for the middle class. For generations, the "American Dream" promised a clear path: work hard, and you could afford a home, send your kids to college, provide decent healthcare, and retire with dignity. It was the aspirational contract of a thriving society. But since the early 1980s, this dream hasn't merely faded; it has been methodically dismantled.

The Housing Crisis: The Ultimate Betrayal of the American Dream

The 2008 financial crisis wasn't a random event; it was the inevitable collapse of the financial casino we discussed in Chapter 1. The architects of modern Wall Street, empowered by deregulation, created complex, high-risk financial products built on a foundation of subprime mortgages—loans given to people who could not afford them. The system, in its unquenchable greed, gave people a false sense of security, selling them the promise of a home and a better life, even as it was setting them up for a fall. When the housing bubble burst, it was not the rich who paid the price. Millions of ordinary Americans lost their homes, their savings, and their credit, while the banks and corporations that orchestrated the scheme were bailed out with the very lifeblood of the working American. Their hard-earned money and their stolen future were sacrificed to save the very system that swindled them.

The question we must all ask is this: Why was bailing out the banks the only solution on the table? Why did the government not offer that same lifeline directly to the millions of Americans on the brink of losing their homes? This is a horrific revelation that unveils a menacing twist—it was a bipartisan political alignment to protect its own diabolical engineers, at the expense of American taxpayers. The betrayal, however, did not end there. After the crisis, the system unleashed a new weapon: corporate landlords. While many families were losing their homes to foreclosure, private equity firms like Blackstone began buying up tens of thousands of single-family homes at fire-sale prices. This wasn't a recovery; it was a hostile takeover of the housing market.

This "new Wall Street" saw housing not as a home, but as a profit-generating asset. They used sophisticated technology and data analytics to optimize their portfolios, and their influence in concentrated markets allowed them to drive up rents and reduce inventory for prospective buyers. The data is damning: renters with corporate landlords often face more frequent and aggressive rent hikes than those with private landlords. The American Dream of homeownership is now a distant fantasy, and a significant portion of the population is trapped in a new kind of servitude—the perpetual rent cycle.

This sobering truth hits hardest when you realize the sheer, physical and emotional toll of this reality. My father, who spent his last years in a nursing home, out of touch with the world, would often ask me, "Did you buy a house yet?" I would simply laugh, trying to make a joke of a tragic reality. His question, born of a past where hard work could lead to homeownership, was a poignant reminder of a dream that no longer exists for so many. The mounting student and credit card debt from my own life, coupled with the instability of my career, hindered me from ever becoming a first-time home buyer. My career was a constant yo-yo of success and struggle. Post-bachelor's degree, I landed a good sales job in the early 2000s, but lost it in 2006. After two years out of work—a period that included the 2008 financial crisis—I took a job for just $9 an hour. It was a time of immense struggle for me and my mother, and the thought of digging myself out of that hole felt impossible.

I eventually landed a job and relocated to California to be closer to my father. My income in the high-cost Bay Area was under $60,000, below industry standard for the time. Finding a place to live was an immediate impossibility. In a final act of desperation and defiance, I accepted the incredible kindness of an elderly friend of the family, who was a resident in a nursing home south of San Francisco. For six months, I slept on an air mattress in her living room in a tiny, one-bedroom condo. Every day after work, I had to be snuck past the security guards by her and her friends—a group of nursing home residents in their late seventies—just to have a place to lay my head. There were even times when I had to crawl through a window because they couldn't sneak me in or her door was locked. This wasn't a choice; it was a brutal, humiliating necessity just to save up enough money for a security deposit and get stabilized again. The system had failed me, and I was relying on the immense humanity of a few strangers to survive.

It was this raw, humiliating act of survival that finally cracked open my understanding of the system. That temporary stability provided only fleeting relief, and the inevitable collapse of that fragile security hit its final breaking point in **2012**.

Looking back, the world I knew *did* end that year, not through a literal Mayan prophecy, but through a calculated, simultaneous systemic collapse designed to extract maximum value across all foundations of life. My struggle wasn't random; it was the direct, humiliating result of a new economic reality where housing became an inaccessible asset, and the burgeoning AI revolution, already known to those in the tech industry, solidified the corporate intent to leave the majority perpetually unprepared. This extraction was total: corporate landlords began a hostile takeover of the housing market to affirm our physical security, while simultaneously, the unchecked growth of social media completed its monetization of our attention and emotional vulnerability which claimed our mental and psychological peace. I realized the system wasn't broken, but designed for this collapse.

I learned the true value of a safety net that the system doesn't want you to have. Having saved enough money, I was extremely lucky to secure a rent-controlled apartment, a literal godsend that has provided stability in a market that has seen so many people around me have their rents double, even triple. Historical data shows this is not a new phenomenon. Practices like redlining, which was a federally sanctioned, racist policy of denying home loans to Black communities, systematically stripped wealth from marginalized groups for decades. This historical injustice, coupled with today's corporate takeover of housing reveals that for certain communities, the American Dream was never a possibility. However, a true, and deeply personal act of defiance against a market that seeks to price us out of a fundamental human right was realized when my partner, an African American, was able to secure a separate house for us to call home—a testament to his love, commitment, and our partnership.

The share of American adults living in middle-class households has steadily shrunk from 61% in 1971 to 50% in 2021, not because everyone moved up, but because a significant portion slipped into the lower-income tier while the top soared. Today, homeownership remains an elusive fantasy for many, particularly in booming urban centers, forcing families into endless cycles of renting or soul-crushing commutes. Healthcare, once a right, feels more like a privilege, with soaring premiums, deductibles, and out-of-pocket costs turning a simple illness into a potential financial catastrophe.

The Healthcare Matrix: Where Your Well-being is a Commodity

The "Matrix" isn't just a digital prison of false reality; it's a physical one built of debt and despair. For the majority of Americans, the promise of a functioning healthcare system has been methodically replaced by a cruel business model that profits from sickness, not health. This isn't a flaw in the system; it is the system! Your well-being, your very right to live a life free from agonizing physical pain, is not a human right; it is a commodity to be sold at the highest possible price. This cold calculus creates a chilling two-tiered system. It's a reality I know intimately. For years, I navigated my own chronic health problems which I was only recently able to diagnose and treat. The fear of what would happen if I lost my insurance was a constant, gnawing anxiety. It was a terror that became brutally real during my years of unemployment between 2006 and 2008. The system, through COBRA, offered a continuation of my insurance, but at a cost so astronomical it was designed to be unaffordable, a financial impossibility for someone without a steady paycheck.

The dehumanizing reality of this system hit its absolute low point one afternoon when I developed a painful staph infection. I was in agony, the kind of pain that forces you to grit your teeth and clench your fists just to hold on. A call to the hospital brought a cold, calculated response: I could come in, but the cost would be thousands of dollars—an insurmountable debt. Instead, I found myself waiting for six hours in a local clinic, sweating and trembling with a pain so unbearable I thought I would pass out. My best friend at the time, a man from a life of profound privilege whose family's wealth insulated him from such a reality, sat with me. Watching me suffer, he said something I will never forget: "No one should ever have to go through this." He was right. And in his profound humanity, he offered a truth the system refuses to acknowledge. Not all of the elite are callous; some are, but it is the system itself that lacks a soul and manufactures inhumanity where you are just a number on a balance sheet.

This illness of the system became even more clear to me later, in California. Even in a state with some of the highest living costs in the country, there is a program called Medi-Cal, which was a lifesaver for me in a different time of need. But even this safety net is a subtle trap. To qualify, you must make a poverty-level income, and if you earn just one dollar over the arbitrary monthly threshold, you lose your coverage and are forced to buy expensive, subpar insurance on the state exchange. What parent with children could possibly navigate that, living on the edge, constantly in fear of losing the one thing that keeps their family safe? It is a political strategy that offers just enough of a safety net to keep you from falling entirely, but not enough to ever truly pull you out of the cycle of dependency. The system is working exactly as it was designed, to keep you from ever truly getting ahead. It's a system where compassion is a political talking point, not a core principle, a system that traps you, regardless of which party you support.

The Food System: Hunger by Design

In a world of unimaginable abundance, where corporations hoard obscene profits, the most fundamental human need—food—has become another weapon in the system's arsenal of control. This isn't just about a scarcity of resources; it's about a manufactured scarcity designed to keep a population compliant and perpetually dependent. In between periods of unemployment when I had literally no income, I lived this brutal reality firsthand on food stamps. Just like with unemployment and healthcare, the system offered a semblance of a safety net, but it was a net full of holes. The amount of money you receive is meticulously calculated based on a poverty-level income. And just like Medi-Cal, if you make one dollar over that arbitrary threshold, your monthly benefits can be slashed, leaving you with scraps. In a high-cost state like California, a mere $290 a month was barely cutting it, and for an entire family, the struggle is unimaginably worse. This is a cruel game of inches, a deliberate policy designed to ensure you never get ahead, but simply remain in a state of suspended animation, just enough to survive, but never enough to thrive.

The real malice of this system, however, lies in the narrative it has cultivated. For decades, the political machine has weaponized food stamps, turning them into a punchline and a tool of political division. Conservative parties have relentlessly framed people on government assistance as lazy, as the scums of society, deserving of judgment and suspicion. They are always ready to rip away school lunch programs for children and deny funding to food banks, all while the system they serve continues to put people in this position in the first place. The real question is not why people are on food stamps; it is why a system of such wealth cannot ensure its citizens have enough to eat without having to rely on the stretched-thin generosity of food banks and the perpetual charity of others?

The irony is a dagger to the heart: in a country with some of the wealthiest people in the world, the safety nets aren't designed to empower the poor, but to manage their poverty. The system needs food stamps not because it is kind, but because it is self-serving. It would rather create a fragile, unreliable program that keeps people from starving than fix the grotesque economic design that created their hunger in the first place. This is the ultimate definition of social control—a system that uses the illusion of help to perpetuate the very injustice it claims to solve. The system isn't designed to solve hunger; it's designed to manage the consequences of its own relentless, unquenchable hunger for profit.

This insatiate appetite for money is a universal force, much like Marvel Comics' Galactus, the Devourer of Worlds. He is a cosmic entity that must consume to survive, and Galactus does so without mercy or conscience. This hunger devours the middle class, consumes the poor, and leaves behind a desolate landscape of debt and despair. This is the true nature of our political system: a soulless machine that must destroy civilizations just to feed its own quenchless hunger, and in doing so, it cannibalizes the very people it was supposed to serve.

The Debt Trap and the Relentless Squeeze

The system's insatiable hunger for profit is not some fictional, cosmic force; it is a very real, very personal pressure that manifests in our day-to-day lives. It is the unseen chains of debt that bind us, the slow-moving catastrophe of stagnant wages, and the constant, relentless squeeze that keeps us running faster and faster just to stay in the same place. This is what the system's hunger truly feels like.

This relentless financial squeeze takes an insidious, invisible toll on our mental and emotional well-being. For parents, it cultivates a pervasive, chronic anxiety. The constant worry about making ends meet, keeping jobs secure, and providing for their children transforms daily life into a never-ending battle. Many feel like slaves to the system, forced into longer hours or multiple jobs, sacrificing precious family time, personal passions, and even sleep, all to maintain a fragile equilibrium. The crushing weight of this burden can lead to a profound sense of resignation, extinguishing hope and vitality, as they watch their own dreams shrink.

This stress cascades onto the younger generation, who are growing up with an acute awareness of their parents' struggles and an unsettling uncertainty about their own futures. They inherit not just financial challenges, but a pervasive anxiety about what lies ahead. Graduating into a world reshaped by economic precarity, technological shifts, and unprecedented debt, they often feel a profound lack of control over their own destinies. This breeds fertile ground for a growing mental health crisis among youth—rising rates of anxiety, depression, and a pervasive feeling of helplessness.

Higher education, the supposed gateway to opportunity, is the perfect example of saddling young people with crippling student loans that often dictate their life choices for decades before they even begin. Rising costs coupled with inflation continues to climb at a rate we've never seen in human history, while real wages for the vast majority have barely budged for decades, even as worker productivity has soared. We are, in essence, running faster and faster just to stay in the same place, or often, to fall further behind.

Adding another layer to this systemic manipulation is the pervasive influence of social media. These platforms were initially touted as tools for connection and community, seemingly designed for good. But like so much else in our current landscape, they were rapidly monopolized by corporate greed. Here's the chilling reality: you are not the customer; you are the product. These platforms are meticulously engineered to gather vast amounts of your personal data—your preferences, your behaviors, your vulnerabilities—which is then sold to advertisers. The algorithms are not designed for your well-being; they are designed for maximum engagement and addiction. They hook you with dopamine hits from likes and shares, cultivate a relentless comparison culture with curated, unrealistic portrayals of others' lives, and erode attention spans with an endless scroll of bite-sized content. This fosters a deep emotional and psychological reliance, particularly among the young, who find themselves caught in a constant pursuit of external validation. They become emotionally and psychologically tethered, losing real-world social skills and battling feelings of inadequacy. This, too, is all done by design.

Just as the financial system is engineered to funnel wealth upwards, the digital ecosystem is engineered to monetize your attention and emotional vulnerabilities. It's another subtle but devastating form of control, extracting not just your financial resources, but your mental peace, your time, and your very sense of self for the profit of a select few. It reinforces the grim truth: the systems that govern our lives are often built to maintain power and extract value, regardless of the human cost.

Chapter 4:

A Personal Reckoning: My Story, Your Reality

The chilling statistics and the calculated mechanisms of wealth disparity aren't just abstract concepts. They are the invisible forces shaping our daily realities, dictating our choices, and often, stifling our potential. If you've ever felt a quiet frustration, a sense that the game is rigged, or a nagging question about why you're perpetually on a treadmill, know this: you are not alone. My own journey through this system has been a stark, often painful, education.

I remember when I first joined Facebook, back in 2009. It felt like a fun, harmless tool, a simple way to connect with friends and see what everyone was doing. But as the years blurred into what feels like mere minutes, a sobering truth dawned: I have hardly any genuine relationships with most of the people on my friend list. What was once a tool for connection had become a master of manipulation. I can personally attest to the countless wasted hours I spent scrolling, consuming, and being unproductive. Hours that could have been invested in making wiser financial choices, nurturing my health, or simply focusing on my actual life.

And while accountability for our choices is undeniably part of the human experience, it's crucial to acknowledge: the systems behind these manipulative psychosomatic strategies are so advanced, so far ahead of their time, that they know exactly what they're doing. They leverage our deepest psychological vulnerabilities, turning us into the product, subtly extracting our most precious resource: our attention and time.

My own financial struggles have been a constant companion, a chaotic rhythm of unexpected highs and punishing lows. Caught in this relentless cycle, I fell into the trap of nurturing my immediate wants, as so many of us do, and this pattern often comes at the expense of our long-term financial health. The reason we fall into this trap is simple and malicious: we feel numb. The constant strain of financial struggle creates a pervasive emotional blankness, and in those rare moments of unexpected high, we desperately seek to feel something—anything. The Mastermind of this manipulated construct knows this instinctively, offering an endless buffet of consumption as a way to dull the numbness, even as it quietly erodes our foundation and keeps us trapped in the very cycle we're trying to escape.

As a huge fan of 1980s pop culture—Voltron, Transformers, G.I. Joe, He-Man—these cartoons served as a comfort during a challenging childhood, a sort of balm for a broken home. As an adult, I've found myself collecting tangible representations of that nostalgia. And while these pieces bring a certain joy, deep down I know it wasn't financially prudent. That money could have been invested, compounding over the years, providing substantial gains that would offer real security, not just fleeting comfort. This isn't just my story; it's a revolving door many fall into, seeking solace in consumption while the system quietly drains their true potential for wealth.

But the system's manipulation doesn't just manifest in our personal coping mechanisms; it is also present in the soul-sucking propaganda it uses to maintain control. I was recently required to attend a company event where they handed out "achievement awards." On the surface, it was a celebration of employees going "above and beyond." But the truth was, the speeches all sounded the same—a series of platitudes about commitment, enthusiasm, and exemplifying the company's mission. I looked at the awardees, and in their perfectly polished smiles, a cold dread set in: they were not being honored for their humanity, but for their flawless willingness to be the perfect cogs in the corporate machine—mindless automatons. When one of my former managers, who shall not be named, a person who weaponized process and micromanagement, received their award, it sounded less like a tribute to a leader and more like a perverse celebration of a slave driver—a person who was rewarded not for humanity, but for ruthless efficiency in upholding the system's demands.

The event was a perfect, sickening example of the corporate nonsense we're all forced to endure. The HR manager, with their sidekick nodding along, talked about globalization, values, and prioritizing internal candidates for growth. It was all a carefully curated narrative, an empty soup of jargon designed to distract us from the grim reality that our true worth is measured in productivity, not in our well-being. This relentless push for "growth, growth, growth" is not about a person's individual liberation; it's about feeding a gluttonous beast that can never be satisfied. It's a game I would never win because I refuse to be the kind of person who deserves that award—a puppet who performs for a pat on the head while the system quietly tightens its grip.

The older I get, the more acutely I feel the cosmic hunger of the corporate machine draining my very life force. I am no longer an employee; I am a nutrient, a temporary source of fuel for a parasite that will never be satisfied. I am required to exchange my time, my soul, for a designated amount of money that an employer, not me, deems worthy. This feeling is not accidental; it's a direct consequence of how our wages have been deliberately suppressed, even as productivity has soared since the birth of the Dow Jones. Wages have barely budged against the rising tide of inflation. This isn't a market anomaly; it's a systemic decision, designed to keep labor costs low and profits sky-high for the ultra-wealthy.

Consider the minimum wage: a foundational protection meant to ensure a living wage. Why, since its very inception, wasn't it federally mandated to increase annually with inflation? Corporate structures are built on budgets; they could easily accommodate these adjustments. Why, then, did the federal government shift the responsibility to individual states, or even cities, creating a chaotic patchwork where millions struggle to survive on stagnant wages? And how can political institutions, with all the data and economic models at their disposal for decades, justify these decisions? Their perennial excuse—that companies would be overburdened—rings hollow when these very corporations have consistently refused to pay fair market value for labor for over 40 years, all while CEOs rake in hundreds of times their average worker's pay.

My personal experience with student loan debt is another stark testament to this deliberate subjugation. Despite securing a scholarship, I graduated with an $80,000 debt for a Master's degree, with no clear path to repayment and an interest rate that feels predatory. Why is the government charging exasperated amounts of interest on these loans? Why are the institutions, and the government that oversees them, so unsympathetic to the crushing burden on millions of students? The Department of Education has never mandated that universities ensure their graduates find jobs, effectively allowing degrees to become, in many cases, literally worthless—massive financial liabilities with uncertain returns. Why would a company demand a Bachelor's or Master's degree for a position, yet refuse to pay the fair market value for that advanced education? It all funnels back to the same conclusion: systemic manipulation to keep us poor. Our time, our comfort, our aspirations, our education—all are leveraged against us, ensuring that the "bigger number" remains perpetually out of reach for the many, reinforcing the cycle of struggle and limited freedom.

Now in my late 40s, after three decades in the workforce, the constant grind of waking up by 8 AM for another day feels like an act of sheer will. My body is exhausted, physiologically drained, requiring nearly an entire day of recovery after my first day off. This leaves me with just one true day of rest, a mere fleeting moment before the cycle begins anew. And to think of Social Security, if it even exists by the time I turn 67, promising little more than poverty-level existence, is to stare into a void of despair.

Every morning, I tell myself, "You just got to do it. You just got to keep going." It reminds me of my dog, Bowie. After he went blind, he became reliant on me to guide him. Despite his age and blindness, he never complained, never balked. He just did it. Bionic Bowie, we called him. He was resilient, loyal, a pure life force. I loved that dog. I had empathy for that dog. He was my lifeblood. He was family, and while my partner provided crucial financial support for his care, I was his primary caregiver, and together we put the needs of our family before our own! But what about the corporations? The entities demanding this relentless sacrifice of our time, our energy, our very essence? They are so utterly separated from their workers, viewing them as mere expendable units—almost like cells in a body, performing tasks without personal consideration.

This cold, calculated transaction was not just an abstract theory; it was a brutal reality I was about to face firsthand. As a top-performing inside sales advisor, I had inadvertently become a threat to the system. With consistent performance reviews that "Exceeded Expectations" and constantly remaining on top of leaderboards driven by my human-centric, consultative approach, I was suddenly placed on a 30-day performance probation with the threat of immediate termination if I did not comply. There was no warning and I had no prior record. The charge? That I violated a fabricated, verbal-only policy that was the polar opposite of my style—a policy that demanded I get customer data before I could even ask why the customer was calling in the first place.

The punishment was so out of context and so disproportionate to a policy that wasn't even in writing, that it was a complete lie, a pretext designed to undermine my professional integrity and threaten my livelihood. My gut, however, told me a different truth: this was a direct attack on my authentic behavior that exposed the system's profound lack of humanity.

In reality, the policy was a cynical gambit to inflate our call volume. By preventing any legitimate, trust-building conversations in favor of a mechanical, rigid, robotic interaction, it became impossible to verify if a lead was legitimate or just window-shopping, which inevitably lowered our conversion rates. The ultimate goal was to pay reduced commissions, a financial liability that executive leadership viewed as a "generous" act to be managed, not as a reward for a successful sales team. My manager, in fact, had told us multiple stories about constantly fighting with executive leadership to ensure we were even paid our commissions. This exposed a system that viewed a successful sales team as an expense, not an asset.

This was the moment I realized the true nature of the corporate machine: my success was a threat to its very existence. My dedication to being authentic, to connecting with customers on a human level, was at odds with a system that demanded robotic, metric-driven conformity. I was being punished not for a lack of performance, but for the unforgivable crime of being so good that I exposed the bogus, anti-human policy at the heart of the corporate machine. The betrayal was not a revelation; it was a physical blow, a gut-wrenching reality that left no room for doubt: there is no empathy, no love, no reciprocal care. It's a brutal calculus where your exhaustion, your limited recovery time, your bleak retirement prospects are simply externalities, unmeasured and unvalued. We are forced to "just do it" because we have no other choice in a system designed to deny us one.

Here's the critical insight I wish I'd had when I was younger: as a teenager or someone in your early twenties, it's almost impossible to truly grasp what life will feel like in your later years with a career of nearly 30 years under your belt, and you're still fighting the corporate bureaucracy, the cynical behaviors, and the underlying, subconscious knowledge that your job isn't truly secure. It takes an unimaginable toll on the mind and body; waking up is almost unbearable some days, and you realize that your youth, your prime, was squandered on a system that was designed to make you feel like a slave. In a world increasingly designed to fragment your attention and diminish your mental clarity, your youth is your prime. This is the very reason for this book: to illuminate these realities now, to provide the insights and tools that most of us learned far too late, so that you can take charge of your own life and navigate these treacherous waters with foresight and power.

Chapter 5:

The Call to Action: The Way Forward

We've peeled back the layers of the system, explored the cold statistics, and confronted the raw, personal toll it takes. We've seen how wealth disparity isn't accidental, how corporate greed reshapes our very well-being, and how society is designed to keep the majority perpetually striving, perpetually dependent. It paints a bleak picture, a relentless cycle that, for many, amounts to a state of modern economic servitude. Not with chains of iron, but with invisible bonds of student loan debt, the relentless grip of credit card balances, the exhaustion of trading precious time for stagnant wages, and the insidious drain on our mental and physical health. It's a servitude to the job, to the debt, to the system itself.

And for the first time in our modern history, we face a forever changing, pivotal moment in the technological revolution. Corporations and economic pundits readily announce that Artificial Intelligence is coming for your jobs, starting with the very entry-level positions that traditionally serve as stepping stones into the workforce. The system, as we've discussed, has fundamentally failed to prepare you for this seismic shift, leaving a generation ill-equipped to find initial stability, let alone thrive.

But here is where the despair ends and empowerment begins. This book isn't just about identifying the problem; it's about providing a roadmap to escape this modern form of servitude. It's about giving you the foresight, the tools, and the strategies to navigate this complex new world on your terms.

The truth about AI is not that it's simply coming for your jobs; it's that it's fundamentally reshaping the way life, work, and wealth will exist. This is not a force to be feared, but a powerful lever to be mastered. While the status quo wants you to remain at its mercy, paralyzed by fear and unpreparedness, this book will give you the precise tools to understand and leverage the power of AI to shape your future. It will help you rethink how you look at this technology, turning it from a perceived threat into your greatest asset.

This isn't about accepting your fate. It's about refusing to be a slave to the default path. It's about reclaiming your power, understanding the hidden rules, and building a foundation that ensures your financial well-being and allows you to unlock your true potential. The glimmer of hope lies in awareness, preparation, and the courage to forge a new path. The choice to break free, to claim a bigger number in your account and, more importantly, a better, more liberated life, starts now.

Part II:

The Factory Floor Mindset

Chapter 6:

The AI Tsunami – Navigating the New Economic Realities

Beyond the Hype and Fear: What AI Really Is (and Isn't)

When you hear "AI," what's the first image that springs to mind? For many, it's Hollywood's version: a Terminator, Skynet taking over, or a sentient robot demanding rights. The headlines often don't help, conjuring images of unstoppable, all-knowing machines. It's easy to get caught up in the hype and the fear, feeling like an inevitable force is bearing down on your job, your life, your very future. But let's take a deep breath and clear the smoke. The truth about AI, especially the kind rapidly reshaping our world today, is far less sci-fi and far more practical. It's not about robots being "born" with consciousness, though it might feel that way with how quickly things are changing. Instead, think of AI not as artificial intelligence, but as intelligence evolving. It's a profound leap in how we process information and automate tasks, but it's still fundamentally a tool, powered by patterns it learns from human data.

I'm incredibly fortunate to have a best friend whose husband is an expert in high tech Machine Learning. This personal connection, coupled with my own exposure working at smaller tech startups that utilized machine learning technology, has allowed us to spend countless hours diving deep into this subject, unpacking the real differences between AI, Machine Learning, and the Language Models you hear so much about. He's patiently explained why, at this stage, we're not heading for a Skynet scenario or a robotic uprising. In fact, we've even discussed episodes of Star Trek: The Next Generation, like "The Measure of a Man" (a classic!), where the crew grapples with the very definition of consciousness. As someone who loves Star Trek not just for the space battles and quirky characters, but for its underlying vision of a utopian future—a future where technology and humanity come together, and where greed simply doesn't exist—it's clear to me: if Gene Roddenberry could envision such a future, why can't AI be used for profound good?

So, what exactly is this "intelligence evolving?" At its core, today's dominant AI is primarily about pattern recognition, prediction, and automation.

- **Pattern Recognition**: Imagine a super-powered detective that can sift through mountains of information—far more than any human could in a lifetime—to find hidden connections. It can look at millions of images and learn what a cat looks like, or analyze billions of sentences to understand how words typically fit together.
- **Prediction**: Once it recognizes patterns, it can make educated guesses. This is why Netflix recommends movies you might like, or why your email can filter spam. AI has learned from past data to predict future likelihoods.
- **Automation**: And finally, it can perform tasks based on these patterns and predictions. This might be writing an email for you, generating a summary of a long article, or even designing a basic image based on your description. It automates what was once time-consuming manual or mental labor.

When you hear terms like Machine Learning (ML), think of it as the AI learning from data without being explicitly programmed for every single possibility. It's like teaching a child to recognize different animals by showing them thousands of pictures, rather than writing a detailed rule for every whisker and paw. And Generative AI (GenAI), like the Large Language Models (LLMs) that power tools such as ChatGPT, are incredibly sophisticated versions of this. They've learned from vast amounts of text and code (all created by humans!) to understand language so well that they can generate new text, translate, summarize, and even brainstorm creative ideas, making it seem eerily human-like.

So, why is this happening now? It's not one single breakthrough, but a perfect storm of advancements. We have unprecedented computing power—literally supercomputers in our pockets compared to decades ago. We have vast, unimaginable amounts of data available online for AIs to learn from. And we have smarter, more efficient algorithms—the instructions that tell the AI how to learn and process information. This confluence has allowed AI to move from theoretical concept to practical reality at breathtaking speed.

The key takeaway here is this: AI isn't a mystical entity with its own agenda. It's a sophisticated system built to find patterns, make predictions, and automate. It's a reflection of the data it's fed which, for now, is overwhelmingly human-created. Understanding this foundation is the first step to moving beyond fear and realizing its true potential as a tool for you.

The Great Reshaping: How AI is Changing Work

The headlines scream about job losses, warning of robots replacing human workers in factories and AI models taking over cubicles. It's easy to panic, to imagine a future where the only jobs left are for AI specialists. But that narrative, while dramatically compelling, misses a crucial truth. AI isn't simply "taking" jobs; it's fundamentally reshaping work itself. It's automating tasks, augmenting human capabilities, and, perhaps surprisingly, creating entirely new roles and industries we couldn't have imagined just a few years ago.

Let me give you a personal example. When I was working in tech, we developed a smart waste bin that used a physical AI machine learning device to sort all waste at the point of disposal. That meant anything—whether recyclable, compostable, or just plain landfill trash—was identified by an AI camera and sorted within its internal chamber. The custodians responsible for managing these units were initially terrified. They saw the machine as a direct threat designed to replace their entire cleaning crew. This fear was understandable; they lacked the education and the policies to truly grasp what this technology was for. My role became about bridging that gap. I created a comprehensive service and maintenance program, essentially designing a new way for these custodians to interact with the machines. I showed them that this wasn't about taking their jobs, but about enhancing them. The AI wasn't going to sweep floors or clean bathrooms; it was going to make their waste management tasks easier, more efficient, and free them up for higher-value work. Once they understood this, once they were given the tools and the knowledge to work with the technology, we saw an incredible increase in operational efficiency, nearly 70%. Their jobs didn't disappear; they evolved, becoming less about monotonous sorting and more about managing and optimizing smart systems. This "Trash Bot" story isn't unique; it's a microcosm of what's happening across almost every sector.

The most significant immediate impact of AI is the automation of routine, repetitive tasks. Think about roles heavy in data entry, basic customer service, or simple administrative duties. AI excels here because it can process vast amounts of information and follow predefined rules with incredible speed and accuracy, without fatigue or error. For example, AI can analyze spreadsheets, financial records, or scientific data far faster than any human, or chatbots can handle common customer queries, freeing up human agents for more complex, empathetic interactions. Even administrative work like scheduling, document routing, and drafting basic emails can be streamlined by AI.

However, this doesn't always mean outright job elimination. Instead, it often leads to job augmentation, where AI acts as a powerful assistant, making human workers more efficient and effective. Imagine doctors using AI to quickly analyze medical images for anomalies, helping them diagnose diseases faster and more accurately, which then allows them to focus on patient care and complex treatment plans. Or lawyers using AI to sift through thousands of legal documents to find relevant precedents, dramatically reducing research time so they can focus on strategy and courtroom arguments. Even writers, designers, and marketers are now using AI to brainstorm ideas, generate drafts, or create initial visuals, allowing them to focus on the creative direction, storytelling, and nuanced human connection. AI is not replacing these roles; it's acting as a powerful super-tool that elevates human capabilities.

Beyond augmentation, AI is proving to be a powerful job creator. Every major technological revolution, from the Industrial Revolution to the internet age, has destroyed old jobs but generated entirely new ones. AI is no different, and in some cases, it's creating surprising opportunities, even for tasks that sound mundane but are critical to AI's learning process. A perfect example of this was the "Trash Bot" I worked on. To train its AI to identify different materials, we needed human workers to manually label scanned items—identifying a plastic bottle, a paper cup, or food waste. These were new jobs, directly created by the need to "teach" the AI. This extends to a range of emerging roles today: AI trainers and annotators who classify and categorize data to feed and improve AI models; prompt engineers who specialize in crafting the precise questions and instructions to get the best and most relevant outputs from large language models; AI ethicists and auditors who ensure AI systems are fair, unbiased, and used responsibly; and AI integrators and managers who help businesses adopt and implement AI tools.

The key takeaway is this: the nature of work is indeed changing, but it's a profound reshaping, not just an elimination. While some tasks will be automated, human ingenuity, creativity, emotional intelligence, and critical thinking will become even more valuable. The new landscape demands adaptability and a willingness to embrace AI as a partner, not a problem.

Chapter 7:

The Factory Floor Mindset – How Education Conditioned Us for Conformity

We've explored the swirling currents of the AI tsunami and seen how a powerful, invisible system leaves many feeling unprepared. Now, it's time to dig deeper into why this happens. The truth is, the system isn't just failing to prepare us; in many ways, it was never designed to. It was built to condition us for a very specific role, one that served the needs of a past era, but now leaves us vulnerable in a rapidly changing world. This is the **Factory Floor Mindset**, and it's been instilled in us from the moment we first stepped into a classroom.

The Assembly Line Classroom: Industrial Roots of Education

To understand this conditioning, we have to look back to the turn of the 20th century, a time of booming industry and mass production in America. This was the era of giants like Andrew Carnegie in steel, J.P. Morgan in finance, and, perhaps most notably for our discussion, John D. Rockefeller, who virtually monopolized the oil industry with Standard Oil. These industrialists faced a challenge: how to ensure a steady supply of obedient, compliant workers for their sprawling factories, mines, and assembly lines. They needed a workforce that understood hierarchy, followed instructions without question, and would show up on time, day after day, to perform repetitive tasks.

Enter the Prussian education system. Originating in 18th-century Prussia (modern-day Germany), this system was designed to create loyal, disciplined citizens and soldiers for the state. It emphasized rote memorization, strict obedience to authority, standardization, and a clear division between intellectual (for the elite) and vocational (for the masses) tracks. Crucially, it was incredibly efficient at processing large numbers of people through a uniform system. Rockefeller, ever the astute businessman, recognized the profound utility of this model. He wasn't looking for innovators or free thinkers; he was looking for reliable labor for his oil refineries and other ventures. A system that could efficiently churn out a predictable workforce, accustomed to following rules and performing repetitive tasks, was a goldmine. He, along with other powerful industrialists of the time, wielded immense influence over emerging public education policy. They saw schools not as places for enlightenment for all, but as training grounds to cultivate the specific workforce needed for their expanding empires. Through philanthropic foundations and political lobbying, they effectively manipulated the nascent American public school system to adopt principles from this Prussian model. It was a conscious effort to prepare a generation not for intellectual exploration, but for efficient functioning within the industrial machine.

Our schools, to this day, often mirror that industrial assembly line. Think about it: the fixed schedules and ringing bells dictating your movement from class to class, much like a factory shift. Students are processed in batches—everyone of the same age moves through the system together, regardless of individual learning pace, unique talents, or areas of struggle. The standardized curriculum emphasizes the transmission of uniform knowledge, not individual exploration or divergent thought. And the hierarchical structure, with strict teacher-student roles, prioritizes authority and obedience. From elementary school to high school, this environment subtly, yet powerfully, conditions individuals for a life of compliance, order-following, and efficient execution within hierarchical systems. It wasn't about teaching you to innovate or become an entrepreneur; it was about preparing you to be a reliable cog in a much larger machine.

The Cultivation of Compliance: Memorization Over Mastery

Within this assembly line, certain practices became paramount, further cementing the **Factory Floor Mindset**. Rote learning and standardized tests sit at the very heart of this. We spent years memorizing facts, dates, formulas, and definitions, often without truly understanding their deeper meaning or how to apply them creatively. Standardized tests, while efficient for large-scale grading and comparison, rarely measure true critical thinking or problem-solving abilities. They reward conformity to a pre-determined "right answer," inadvertently penalizing anyone who thinks differently or questions the established method. This system actively discourages critical thought. When you're constantly pushed for "the right answer" on a test, there's little reward for questioning how that answer was derived, or why a particular method is preferred. The emphasis shifts from the process of inquiry to the end result of conformity. This subtly trains us to accept information as presented, rather than to analyze, challenge, or synthesize it independently. The pressure to "fit in" only intensifies this. From playground dynamics to classroom expectations, a focus on blending in, on not rocking the boat, can subtly suppress individuality and creative expression.

The Unwritten Curriculum: Missing Manuals for Life

Perhaps the most insidious aspect of the **Factory Floor Mindset** is not what it teaches, but what it deliberately leaves out—the unwritten curriculum of essential life skills. We spent years in school, yet where was the practical financial literacy? We graduated with little to no understanding of basic budgeting, how to invest, the true dangers of debt (especially student loans), how credit works, or strategies for building personal wealth outside of a paycheck. This absence keeps individuals perpetually dependent on employment, because they lack the fundamental knowledge to manage and grow their own resources, or even to understand how financial systems truly operate.

Even more glaringly, the system offers almost no pathways or encouragement for entrepreneurship. The very idea of charting your own economic course, of becoming your own boss, often feels not just alien, but like trying to climb an unclimbable mountain without a map or gear. While the traditional path is meticulously laid out, the path of the entrepreneur is left as a wild, untamed wilderness. There are no standardized tests for innovation, no degrees in bootstrapping, no lessons in navigating the labyrinth of regulations or pivoting when your first idea inevitably crashes. Generations are sent out with the implicit message that the only valid role is to be an employee, not a creator of employment. Yet, despite this systemic neglect, countless individuals around the world still manage to summit these "unclimbable" peaks, driven by sheer will and a relentless grit to pivot through endless obstacles and red tape. And here's a crucial, often overlooked, truth: even those who successfully break free and become their own boss often fall back into the very patterns of the system they escaped, keeping wages stagnant and resist providing benefits for their own employees, perpetuating the cycle of the **Factory Floor Mindset** from a new position of power. Their success often stems not from what they learned in the system, but from what they learned despite it, and tragically, sometimes they apply those same limiting principles to others.

This is the unwavering reality of the system's conditioning: it keeps you locked in the employment cycle, afraid to even consider entrepreneurship. And here's why that cycle is so difficult to break: The absence of education on workers' rights, labor history, or the power of collective action like unions leaves individuals disempowered. They're often unaware of their collective strength or how the economic system has been shaped to benefit some at the expense of others. This lack of awareness keeps the workforce fragmented and less able to advocate for its own interests, perpetuating the cycles of suppressed wages and increasing wealth disparity we discussed earlier.

The At-Will Doctrine: The Law of the Corporate Cage

The legal mechanism that exploits this very disempowerment has been waiting in the playbook for centuries: the at-will doctrine. "At-will" is not a modern innovation; it is a legal relic that was given new life by the masterminds of the system. Its origins trace back to the 19th century, a time when titans like Rockefeller and Carnegie were building empires on the backs of an unprotected workforce. It was a legal theory, championed by scholars like Horace Gray Wood, that served to liberate employers from the old common-law tradition of having to show "just cause" to fire a worker. The old common law tradition was a legacy of English law that presumed employment was a yearly contract, which provided workers with some stability and required an employer to have a valid reason for termination. This tradition was seen as a hindrance to the titans of the Industrial Revolution who wanted complete control over their workforce. It was a time when employers were increasingly seen as the drivers of economic progress, and legal scholars like Horace Gray Wood championed a new legal theory—freedom of contract—that argued employers and employees should be free to terminate a relationship at any time for any reason. This legal principle, which was never voted on or passed into law, instead became the standard practice across the country as judges began to consistently rule in its favor. This was a quiet, subtle victory for the industrial class that was largely ignored by the public, precisely because it was not a public vote or a new law, but a gradual shift in the courts that gave employers more control over their workers. For workers, the concept of "at-will" was not part of their vocabulary, but they felt its effects firsthand. Their fight for job security, fair wages, and protections through unions was, in effect, a protest against a system that was increasingly treating them as disposable.

This movement successfully kept the doctrine in check for decades, but it was a victory that would not last. In the 1970s and '80s, the same masterminds who were deregulating the financial markets, giving us the casino of Wall Street, saw an opportunity to give the at-will doctrine new life. Fueled by economic downturns that made workers desperate for a job and a pro-corporate agenda that was hostile to unions, the masterminds gave the at-will doctrine a new legal push. This philosophy, once again championed in the courts, became a legal weapon perfectly suited for the same era of deregulation and corporate empowerment that created the financial casino. The at-will doctrine was not an accident of law; it was a tool, waiting to be used.

The true horror of this doctrine is how it's used. At-will employment does not legally supersede protection based on age, race, sex, or other protected characteristics. But the system's design is laid bare: it gives you the illusion of protection with one hand while holding a legal ax over your head with the other. Companies can, and will, simply make up a reason to fire you, and the burden is then on you, the employee, to prove that their reasoning was a lie—a legal process that is almost always too costly and too difficult to win. This is why most people, especially the young, do not fully comprehend its power. It's not a law that's easy to see, but a subtle, deceptive legal phrase, and its true power is often ignored. In contrast, employment contracts are the norm in Europe and Canada and a company must provide several months' notice or severance, demonstrating that our system is not a universal truth but a deliberate choice. It is the ultimate tool of control, the legal foundation of the corporate cage that makes a person's employment little more than a fragile, temporary loan of a paycheck from a system that could, and would, take it away at a moment's notice. This is the grim reality of conditional employment, and the lie that is spun to justify it. For me, the first sign of this deep-seated control was in every employment contract I ever signed. I would sit there, with my inner reservations screaming, as I read the subtle statement that says, "This company abides by the at-will doctrine." And while my own employer once tried to sell me the insulting facade that at-will was great because it also gave me the freedom to quit at any time, I saw right through the lie. This terrifying reality of conditional employment is the very epitome of layoffs, and a constant, unspoken fear for anyone trying to build a future. It creates a workforce that is always on edge, always in a state of precariousness, and always in a state of conformity.

In the decades since its resurgence, here is the most disturbing detail of all: on an issue that makes your livelihood a fragile, temporary loan, neither political party has ever made a serious effort to change it. This isn't an oversight; it's a quiet, bipartisan agreement that ensures the system's power remains unchecked, serving a master far more powerful than any politician: greed itself.

The Corporate Shield: The Weaponization of HR

The system has a final trick to ensure compliance: the profession of Human Resources. Human Resources was originally born out of a perceived need to look after the "personnel" within an organization, a seemingly benevolent field focused on employee well-being and managing the human element of a business. But this noble origin is a lie. The system pulled the wool over our eyes, convincing us that HR was there to help us. But a seismic shift in the mid-20th century, particularly after the Civil Rights Act of 1964, would fundamentally alter this relationship, turning HR from a worker-focused advocate to a corporate shield. The act created a new legal frontier for civil rights, but for corporations, it created a new and immense risk—the risk of being sued.

In this new legal landscape, HR's role evolved from a benevolent "personnel manager" to a legal defense force. The human element of the business was no longer a person to be cared for, but a legal liability to be managed. Over the decades that followed, the field became weaponized by the system, building a vast bureaucracy that created a web of policies and procedures. This web was not designed to protect the worker; it was designed to insulate the corporation from lawsuits, to create a paper trail that could be used to justify a termination, and to manage a compliant workforce through the fear of legal retribution. The collective "chill" that people feel when they talk about HR is a subconscious recognition of its true role: it has evolved into a corporate shield, a department designed to protect the company's interests above all else. This weaponization of HR has led to the complete dehumanization of the workforce. Harassment and safety policies, which were created to protect workers, have been repurposed as tools of corporate control, used to manage internal disputes and silence dissent. Employees within HR, often with good intentions, have been conditioned to see their colleagues not as people, but as legal liabilities to be managed and contained. This can manifest in speeches about "globalization" but the message is always the same: constantly grow your skill set, conform to the corporate mission, and you might not be the next one on the chopping block. The globalization of the modern economy only intensified this. As companies expanded across the world, HR's role became about ensuring a standardized, risk-free model of employee management on a massive scale, turning individual human beings into data points to be managed and contained. Part of the management and standardization process is a concept called "Upskilling," which is nothing more than a hollow promise sold as a tool for "growth." In reality, it's a way for the company to shift the burden of professional development onto the employee while giving them no real power. It's also a tool for compliance, and my own experience confirmed this chilling truth: employees who are not compliant or resist these trainings are marked as a potential future liability for the company. They are seen as non-conformists, and in a system built for obedience, conformity is a virtue and dissent is a threat.

The final, brutal irony of this system is that the HR employees who guard the shield are often immune to the very chopping block they manage— their purpose, after all, is to protect the system, and the system, in turn, protects them.

The Golden Handcuffs: The Promise vs. The Reality

For generations, the promise was clear, drilled into us from childhood: "Go to school, get a degree, find a good job, and climb the ladder." This was sold as the infallible pathway to security, the formula for success, the very definition of the American Dream. Yet, for millions today, this promise has morphed into a pair of golden handcuffs, binding them to a system that demands their conformity while offering diminishing returns. The emotional toll of this conditioning is perfectly captured in pop culture.

Think about the song "Manic Monday" by The Bangles. That universal dread of the alarm clock on a Monday morning, the desperate wish for "just one more day" of freedom, the frantic rush to "be at work on time or you won't get paid"—it's an anthem for the Factory Floor Mindset. It powerfully articulates the psychological burden of a system that conditions us to view work not as a fulfilling endeavor, but as an inescapable obligation, where presence trumps passion, and time-stamped attendance is the primary measure of worth.

This pervasive attitude of the ruling class is perfectly encapsulated in the Harry Potter series by figures like Lucius Malfoy, a character steeped in inherited privilege within the Ministry of Magic. Malfoy personifies the deep disdain for those who must toil within the system; we see it in his sneering remarks towards Arthur Weasley, Ron's father, a dedicated Ministry employee perpetually working overtime in the less glamorous departments. Malfoy's contempt for Arthur's meager income and "Muggle-loving" ways isn't just personal animosity; it's the elite's scorn for anyone who doesn't operate within their sphere of inherited wealth and unchecked power. It's the belief that those who work hard but remain financially constrained are simply "lesser," a mentality that reinforces the very class divisions the "Factory Floor" system was designed to maintain. This, in turn, creates a subtle form of servitude, a feeling of being eternally bound, much like the house-elf Dobby. Yet, the profound desire to escape such an existence of forced labor culminates in an explosion of joy upon receiving something as simple as "clothes"—the literal key to his liberation, allowing him to declare, "Dobby is free!" This triumphant moment underscores the universal yearning for true autonomy and the sheer delight found in breaking free from the shackles of servitude.

My own career journey is a stark illustration of this trap. In the first part of my professional life, I was utterly a slave to this mindset, relentlessly trying to climb the corporate ladder. My roles were heavily sales-driven, where I, as a representative, theoretically had control over my income based on the sheer amount of work I put in. I was fiercely dedicated, often approaching sales in a creative, unconventional way that consistently yielded top results—I'm talking number one, or always in the top five, year after year. Yet, despite these undeniable successes, I was constantly scrutinized and, at times, even disciplined for not rigidly conforming to the sales processes outlined by my employers. It made no sense to me then, and it still doesn't, that achieving superior outcomes could be met with punishment simply for not following the exact, prescribed path.

Even back then, though unconsciously, I began to sense I was caught in some kind of Matrix, an invisible system that imposed a ceiling I couldn't break. I would try to climb that ladder, push for management roles, but I was always met with the same dismissive feedback: "You're a great representative, but you'd never be a great manager." The explanations were always vague, the reasoning flimsy. It became clear that being a "great rep" wasn't enough if you weren't also a perfectly compliant one, willing to stifle your own methods for the sake of conforming to an outdated structure.

Seeking a different path, I pivoted, gravitating towards startups where the environment was, at least on the surface, more dynamic. Here, I found opportunities to create new methodologies and brainstorm innovative solutions, momentarily escaping the rigid status quo of traditional Corporate America. But even within these seemingly looser structures, the underlying factory floor mentality, driven by traditional leadership hierarchies, still existed. There was always a ceiling, a subtle yet firm hand guiding me back towards a predefined path, never quite allowing full autonomy.

Adding another layer to this disillusionment, I later pursued a Master's Degree, believing it would be my ticket to higher income and greater opportunities, a solid investment in my future. I remember the immense financial burden, the hope I clung to. Yet, the reality was a brutal awakening: that degree did not yield the results I had hoped for, neither in terms of income nor career advancement. The system that pushed me towards accumulating more education had, in fact, saddled me with more debt, forcing me deeper into the very reliance on a paycheck it claimed to liberate me from.

This isn't just a warning for youth; it's the lived reality for a vast segment of the adult population, a stark testament to how effectively the Factory Floor Mindset has trapped us. We bought into the promise of climbing a ladder that, for many, has either been dismantled, or leads only to a dizzying view of mounting bills and diminishing returns. The illusion of security in a traditional job has eroded, yet people are still conditioned to pursue it, often at the cost of their adaptability, their financial well-being, and their very peace of mind.

Breaking the Mold: Reclaiming Your Power

Understanding how you've been conditioned is the first, crucial step toward liberation. For teenagers and young adults still navigating the traditional education system, it can feel incredibly difficult to imagine breaking free from its patterns. But this chapter isn't about blaming individuals; it's about exposing the systemic design. It's about empowering everyone, regardless of age, to recognize the invisible shackles of the Factory Floor Mindset and to choose a radically different path. It's about consciously unlearning years of conditioning—to question, to adapt, to innovate, and to claim a future where you are the architect, not just a compliant blueprint.

Chapter 8:

The Unprepared Generation: The System's Hidden Playbook

The AI tsunami is no longer a future threat; it is here, and it's promising to reshape life, work, and wealth as we know it. But why, in a world drowning in data, are so many of us—especially the younger generations—so utterly unprepared? The answer is not a lack of effort or intelligence; it's a profound systemic betrayal. The game was rigged long before we were given the rules. This betrayal is most chillingly embodied by a simple truth: they have a playbook, and you don't.

The Criminal Playbook

Nowhere is this truth more exposed than in the brutal reality of insider trading. We are told that such a thing is a crime, a violation of the "free market." Yet, the very act of trading on private, non-public information is a daily reality for the masterminds of our financial system. The most egregious acts are often committed far from the public eye—not because they don't happen, but because they are intricate, complex, and woven into the very fabric of how the system operates, making them incredibly difficult to detect and prosecute. The system is designed to protect itself.

The most prominent cases are often sensationalized, but the truth is, they serve a greater purpose. The infamous case of Martha Stewart, for example, was a public spectacle, and many critics believed that her celebrity status and gender made her an easy target. But the system is not blind to the wealthy; rather, it often targets those who operate too brazenly outside of the established norms, even if they are part of the elite. The massive insider trading rings of men like Michael Milken and Raj Rajaratnam, which involved systemic fraud on a monumental scale, were impossible for the system to ignore. What is a punishable crime for some is simply a different form of the same game for those who write the rules.

The system even has laws designed to, in theory, prevent this. Antitrust laws are supposed to be the great equalizer, a legal guardrail meant to prevent corporate giants from amassing so much power that they can rig the entire market. But the masterminds of the system have rendered these laws toothless, using their political influence to ensure that a few corporations can consolidate their power and control every aspect of an industry, creating a market of monoliths that we are all forced to navigate.

The Illusion of Choice: The Market of the Monoliths

This consolidation of power is not a new phenomenon, but it is one that has only intensified since the golden age. While the titans of the golden age built their empires through cutthroat tactics, they are now dwarfed by the corporate monoliths of today. What was once the dream of building an empire has become a sinister reality of market control. A handful of companies, many of which you know and use every day, have consolidated their power to an unimaginable degree, creating the illusion of a diverse, competitive marketplace.

Look at your grocery store shelves. You see dozens of different brands of cookies, cereal, bread, soda, toothpaste, shampoo and body wash and so forth. But look closer, and you'll find that all those brands are owned by a mere handful of mega-corporations like Procter & Gamble, Unilever, Coca-Cola, and PepsiCo. We are given the illusion of choice, while the reality is that the control over what we eat, drink, and use is in the hands of a few. This consolidation of power is the embodiment of what antitrust laws were supposed to prevent. They are the market's "sovereign overlords," the modern-day trusts that were outlawed in theory but now dominate our lives in practice.

This problem has only intensified with the rise of the digital age. A handful of tech giants—Google, Apple, Amazon, and Meta—now control our information, our communication, and our commerce. They have created a new, digital form of market concentration that is more pervasive and harder to fight than anything the world has ever seen. What you buy, what you read, and who you communicate with are all filtered through their platforms, giving them a level of power and control that is terrifying in its scope. This is not a competitive marketplace; it is a meticulously engineered Monopoly, where the rules are written to ensure that the house always wins.

The Political Playbook

So who exactly is writing these rules? The collusion between Wall Street and the political establishment is so deeply woven that they are practically a single entity. The architects of this system didn't just reward themselves with money; they rewarded each other with the ultimate currency of power: inside information. A recent article from a nonprofit group exposes the ongoing conflicts of interest in Congress, citing how stock trading continues to raise concerns. The fact that a new ethics bill is even being proposed to address this issue is the ultimate indictment of a broken system. This isn't a one-party problem; politicians from both sides of the aisle, from Nancy Pelosi to Marjorie Taylor Greene, have been accused of the very same thing. This corruption is not a failure of character; it is a profound perversion of a system where our elected officials are not serving the people, but the financial interests that put them there.

The sovereign corruption of this system is so profound, it has become the very DNA of the Oval Office. Isn't it ironic: a person elected to serve the people often uses that power to cash in with the very same financial masters they were supposed to regulate. We've seen it firsthand, from presidents who leave office and immediately earn obscene, multi-million dollar paydays from the very Wall Street firms they once claimed to regulate, to others whose vast business empires create a constant, unspoken conflict of interest. The political system and the financial system aren't just connected; they're two sides of the same corrupt coin. Their interests are one, and they are served at the expense of the American people, and the law becomes a tool to be navigated for personal gain.

The Comedy of Cruelty

The system's cruelty is so blatant, so utterly detached from human reality, that for the elite, it has become a dark comedy. Imagine a famous Saturday Night Live skit where Alec Baldwin, playing a smug, wealthy banker, is shown a video of an average American family losing everything in a foreclosure. The banker and his friends simply look at the screen and erupt into a fit of laughter, finding the family's pain to be the punchline of a sick joke. That skit is not just a parody; it is a mirror reflecting reality.

Shows like In Living Color and Mad TV, through their raw and unapologetic humor, often targeted the brutal truth of this system. In Living Color's "The Homey the Clown" skits, for instance, used comedy to expose social and racial inequality with a raw honesty that felt like a slap in the face. Similarly, Mad TV's sketches, like "Miss Swan" or those targeting corporate culture, used a very specific, cringe-inducing humor to highlight the absurdity of a system that so actively works against a large portion of the population. The fact that the suffering of the poor and middle class is either the subject of ridicule or a footnote in a financial report highlights an unconscionable truth about the elite. They don't just see the world differently; they see it as a game, and your struggle is simply the cost of admission.

Chapter 9:

The Echo of Early Choices – Financial Foundations for Life

The shadow cast by our early choices, or lack thereof, can stretch across an entire lifetime, particularly when it comes to money. We've discussed how the traditional education system often leaves us unprepared for the complexities of the modern world, conditioning us for conformity rather than financial mastery. For many, myself included, this lack of foresight and practical financial guidance led to a series of choices that echoed through decades, shaping paths in ways we never intended.

I can be the first to admit that I made some truly regrettable financial choices early in my career. Fresh out of school, despite having a full-time job and a successful sales career, I simply did not recognize the immense, compounding value of contributing to a retirement account that my employer offered. It was free money, a golden opportunity for my future self, yet I let it slip by, unconcerned. Worse still, I was a member of the union at that employer, and while I dutifully paid my union dues, I remained blissfully unaware of the pension plan they also provided. It wasn't until much later that I even saw it show up in my retirement account. While the pension was only projected to give me a modest $200 a month upon retirement, I made another crucial misjudgment: I decided it would be a better investment to cash it out entirely. My rationale, born of desperation and a potent cocktail of hype, was that an exploding market—specifically, marijuana stocks—would be my quick ticket to breaking free from the shackles of employment. I saw the promise of immediate wealth, a fast track out of the Matrix. What I tragically lacked was due diligence, proper research into the inherent risks, or the wisdom to consult a qualified financial advisor. The result? I lost my entire pension. That significant sum, meant to provide a sliver of security in my golden years, vanished due to an ill-informed gamble.

My father, bless his foresight, had always warned me to avoid credit card debt, and for the longest time, I managed to steer clear. But then, one single mistake changed everything. Driven by a nascent, misguided ambition, I borrowed money against my credit card to bet on an oil stock that, as you might guess, ended up going under. I saw it as another chance to get ahead, to beat the "system." In a moment of profound irrationality, born from the same desperate hunger for liberation, I repeated my mistake. I made an uncalculated gamble on a market that, for the uninitiated, is not a fair game—it's a rigged table where everyone else has a secret cheat sheet.

The consequences of these missteps have been relentless. For years, I've been in a constant struggle, trying to climb out of that financial hole. Every time I thought I was making progress, inflation seemed to rise up, beating me down again, eroding any gains. It has significantly hindered my ability to contribute meaningfully to my retirement at an older age. And here I am, almost fifty, feeling as if I'm starting from scratch. It's a frustrating, humbling reality born from early choices, or rather, the lack of informed choices, driven by a system that prioritizes conformity over competence in personal finance.

The long-term impact of these early financial decisions, or the lack of them, cannot be overstated. When we don't start investing young, when we fail to understand the power of compound interest, or when we make impulsive, uninformed moves in complex financial markets, the cost isn't just immediate; it's exponential. A missed dollar invested at 20 is worth exponentially more than a dollar invested at 40 or 50. The ripple effects of debt, missed opportunities, and financial illiteracy create a perpetual struggle for stability, making it incredibly difficult to truly break free from the "Factory Floor Mindset" that demands constant labor just to stay afloat.

This struggle is exacerbated by a societal design that actively pushes us towards financial vulnerability. Why did we become a credit-based society in the first place? And why are we so strongly encouraged to spend money we don't physically possess in a bank account? This shift wasn't accidental. After World War II, as consumer culture boomed, financial institutions and industrial leaders saw an unprecedented opportunity to stimulate economic growth - and their own profits - by fostering continuous consumption. Credit cards, once a novelty, became a ubiquitous powerful tool for accelerating spending by leveraging future earnings. They created a cycle of immediate gratification and deferred payment, benefiting banks through interest and fees, and corporations through increased sales.

At the heart of this credit-based system is the FICO score, a seemingly innocuous three-digit number that holds immense power over your life. FICO, or Fair Isaac Corporation, developed this score to summarize your credit risk based on your payment history, the amount of debt you carry, the length of your credit history, and more. It dictates your borrowing power - whether you can get a loan for a car, a mortgage for a home, or even secure certain jobs. A low score can quite literally block your access to the very opportunities necessary to build financial stability. The institutions that control this system knew precisely what they were doing when they put these powerful tools into the hands of a population that was largely financially uneducated. Tying back to the Factory Floor Mindset, Rockefeller and other industrial leaders weren't just creating compliant workers; they were fostering a compliant consumer base that would borrow, spend, and remain perpetually engaged in the economic system they controlled. They knew the psychological impact of easy credit on individuals not trained in financial discipline, and how it would keep them running on the hamster wheel, endlessly trying to catch up.

Indeed, as a reputable financial TV personality once succinctly put it, the truly wealthy often laugh at FICO scores. For them, it's a metric of control, a lever designed to manage the borrowing of the masses, not a personal concern. Their vast assets, established relationships with private bankers, and the ability to leverage collateral or direct cash flow means they operate in a financial stratosphere where a FICO score is largely irrelevant. It's a cruel irony: the system imposes this rigid, unforgiving scorecard on those with the least financial resilience, while those at the pinnacle of economic power simply bypass it entirely.

The pervasive nature of this system is perhaps best illustrated by a game almost everyone knows: Monopoly. What many don't realize is its true origin.

Originally conceived in 1903 by Lizzie Magie as "The Landlord's Game," it was intentionally designed to expose the economic perils of land monopolization and the single tax theory. Her purpose was to vividly demonstrate how a system where a few accumulate all the property inevitably leads to the impoverishment of the many. Yet, ironically, this didactic tool was re-packaged and popularized as Monopoly, a game that celebrates the ruthless accumulation of wealth and the inevitable bankruptcy of those who fall behind. It's a game where the rules, once understood, reveal a built-in advantage for those who start ahead, and a predetermined outcome for those who don't. This subtle, yet profound, manipulation of a simple board game mirrors the larger financial structures that govern our lives.

This score, therefore, is not a universal measure of financial health, but rather a tool of the Matrix, designed to perpetuate a cycle of dependence and limit the upward mobility of those already struggling to escape the factory floor. The consequences of early financial decisions, especially those involving credit, can ripple for decades, creating a low FICO score that acts as a gatekeeper, hindering your ability to secure loans, better interest rates, and ultimately, true financial freedom.

This personal journey, however, led me to a profound realization. This very struggle, this awareness of being caught in an economic Matrix, fueled my drive to find a different path. It was through my work at startups, observing the rapid evolution of technology, that I began to see AI not as a threat, but as a powerful, liberating tool. I realized that the same information asymmetry we discussed earlier - where the wealthy leverage superior knowledge - could be profoundly disrupted. Imagine a world where young people, no longer shackled by the shortcomings of traditional financial education, can consult with AI. Instead of blindly trusting vague advice or making uninformed gambles, AI can analyze the Dow Jones Industrial Average, dissect every mutual fund, scrutinize every ETF, and project potential returns. It can identify emerging industries, spot hidden opportunities, and even help craft personalized investment strategies that were once the exclusive domain of the rich, guarded by high-fee financial advisors. And I can personally attest to AI's power here. In my current state, facing the daunting prospect of starting my retirement savings from scratch, consulting with AI regarding the plans available in my 401k has been transformative. Its recommendations have put me back on track, yielding higher returns than anything I could have ever hoped for or achieved through conventional means. This isn't just about recovering from past mistakes; it's about proactively building a solid financial foundation, turning the tables on the rigged game, and ensuring that no one has to start from scratch in their later years because of early, uneducated choices. This is where AI truly becomes a catalyst for financial freedom, a powerful co-pilot in navigating the complexities of wealth building for everyone, potentially democratizing the role of a financial advisor.

Part III:

Your Blueprint for Liberation

Chapter 10:

Your AI Toolkit: A Roadmap to Empowerment

Now that we've fully explored the systemic cage and understood our own conditioning, the crucial pivot begins here, with your **AI toolkit**. This isn't a book about fear; it's a manual for **empowerment**. This isn't about becoming a coding wizard overnight or mastering complex algorithms; it's about a fundamental **shift in your perspective** on the future, armed with the most powerful tools available.

The Mindset Shift: Curiosity, Courage, and the Quest for Knowledge

The first and most vital tool in your AI toolkit isn't a piece of software or a complex algorithm; it's your **mindset**. For too long, the system has conditioned us for compliance, to follow instructions, and to outsource our critical thinking. Now, it's time to reclaim your **intellectual autonomy**.

This begins with **curiosity**. True curiosity is a rebellious act in a world designed to spoon-feed you information. It's about asking "why" and "how," venturing beyond the headlines, and exploring the unknown. Embrace the discomfort that comes with learning something new, with stretching your mind beyond its familiar boundaries. It's in this space that true understanding and innovation happen.

And central to this sense of inquiry is **reading**. Think of **reading** as your **golden wand, your superpower** in a world awash with fleeting trends. Remember Hermione Granger from *Harry Potter*? Socially a little awkward at times, perhaps, but undeniably one of the smartest witches of her generation. Why? Because she read. She wasn't talking about skimming headlines or scrolling social media; she devoured weighty tomes, seeking knowledge, meticulously piecing together clues, and consistently finding solutions others missed because they simply hadn't done the reading. This is your cue. Don't limit yourself to what's easy or what confirms your existing beliefs. Dive into books that pique your interest, yes, but also those that challenge you, that open up new worlds of thought, and ignite new sparks of inquiry. Your ability to absorb and synthesize information—to weave disparate facts into a cohesive understanding—will be your greatest asset.

We are in a generation that is already digitally native, comfortable with technology woven into every fabric of life. The mindset shift here isn't about adopting technology, but about consciously taking the leap to use AI as an individual tool for creation and income, rather than just consumption. Stop thinking of AI as merely a search engine or a social media feed. Instead, imagine it as a brilliant, tireless co-pilot. Your challenge is to ask: "What can you do for me? How can you help me earn income, learn a new skill, or automate a tedious task so I can focus on building my own future?". This isn't just about passive learning; it's about active experimentation and pushing the boundaries of what's possible for you. You already know how to wield a smartphone; now it's time to realize that AI is just the next evolution of that powerful capability, waiting for your instruction.

Key Skills to Develop: Beyond the Classroom

The traditional education system might not be teaching you these, but they are the bedrock for thriving in an AI-powered world. These aren't just about technical prowess; they're about how you interact with and harness intelligent tools.

- **Prompt Engineering and Effective Communication with AI.** This is less about writing complex code and more about clear, precise thinking. AI models are incredibly powerful, but they are only as good as the instructions you give them. Learning to craft precise, detailed, and strategic **"prompts"** is a new, high-value skill. It's about translating your human intent into language AI understands, ensuring you get valuable, relevant output instead of generic noise. Think of it as learning how to truly speak to your digital co-pilot to unlock its full potential, guiding it to produce exactly what you need. I can personally attest to the power of this skill. When I was actively looking for a new job, I used AI like Gemini to modify my resume, helping me articulate my skills in ways I hadn't considered. It also became an incredible individual partner for brainstorming ideas on how to monetize my skills, essentially acting as a private business consultant. This wasn't just about typing commands; it felt like a focused conversation with an incredibly knowledgeable assistant, always pushing towards a clear direction and outcome.

- **Critical Thinking and Verification. AI generates; humans verify**. While AI can produce text, images, or code at lightning speed, its outputs aren't always perfect or accurate. Developing sharp critical thinking skills means you can evaluate AI-generated content, fact-check it against reliable sources, identify potential biases, and refine it to meet your specific needs. **AI gives you the data, but your critical thinking is the human filter that judges, verifies, and applies that data.** This isn't just about spotting errors; it's about ensuring that what AI produces is truly useful, ethical, and aligned with your goals. You're the editor, the curator, the final arbiter of truth and quality.

- **Adaptability and Continuous Learning.** The pace of change will only accelerate, meaning the most valuable asset in the AI era isn't a fixed skill set, but the inherent capacity to learn, unlearn, and relearn constantly. This means embracing a growth perspective, viewing new technologies not as threats, but as new landscapes to explore and master. It requires actively seeking out new courses, engaging with online tutorials, joining communities of learners, and, most importantly, simply experimenting on your own. Your education is now a lifelong project, and you are its primary architect.

- **Domain Expertise combined with AI Fluency.** The most powerful combination in the coming years will be deep knowledge in a specific field—combined with a strong understanding of how to utilize AI tools within that field. This allows you to become a "super-performer," someone who can harness AI to achieve results far beyond what human effort alone could accomplish. You become the specialized guide who can instruct the AI precisely, interpret its results, and apply them meaningfully within your chosen domain.

Practical Applications: Putting Al to Work for You

This isn't theory; it's immediate action. You can start integrating Al into your life today to boost your productivity, unleash your creativity, and accelerate your earning potential. Think of Al as your **personal, always-on assistant** for a vast array of tasks:

- If you need to generate content, whether it's a compelling email, a draft for a blog post, or ideas for a presentation, Al can provide a strong starting point, saving you hours of staring at a blank page.
- For information synthesis, if you have a lengthy report, an academic paper, or even a dense book you need to understand quickly, Al can summarize complex information into digestible points, highlighting key takeaways in minutes.
- Want to learn a new skill? Al can act as a personalized tutor, explaining concepts, providing examples, generating practice problems, and even helping you brainstorm ways to learn more effectively.
- For creative brainstorming, when you're stuck on a design idea, a marketing slogan, a name for your new business, or a plot twist for a story, Al can generate countless variations and fresh perspectives to spark your creativity.
- Even for basic research and analysis, Al can quickly pull together information, identify trends in data, or help structure an argument for a project, though remember, human verification is always the essential final step.

By adopting this proactive mindset, by diligently cultivating these key skills, and by actively experimenting with Al's practical applications, you won't just survive the Al tsunami - you'll learn to surf it. You'll transform what others fear into a powerful wave that carries you towards a future of your own design, one where your human ingenuity is amplified by intelligent tools.

Chapter 11:

The Human Advantage – Unlocking Your True Potential with AI

As we stand at the precipice of an AI-driven future, it's easy to feel dwarfed by the sheer processing power, data analysis capabilities, and tireless efficiency of artificial intelligence. But despite its awe-inspiring capabilities, AI is fundamentally a machine. It lacks something profoundly essential, something that makes you uniquely powerful: your humanity.

Beyond the Machine - What Makes Us Uniquely Human

AI can sift through mountains of information in seconds, identify patterns invisible to the human eye, and execute complex calculations without error. At our core, to err is human. We are born inherently flawed, not as perfect constructs, but as blank slates, like sponges ready to be molded. This very malleability, this capacity for mistakes and the subsequent growth from them, has historically been both our greatest vulnerability and our most profound strength. It's the vulnerability that those in power have exploited, shaping us through systems designed for conformity. Yet, it is also the wellspring of our boundless capacity for learning, adaptation, and truly unique contributions that AI simply cannot replicate.

Consider the beloved character of Data from *Star Trek: The Next Generation*. A highly advanced android with unmatched processing speed and knowledge, Data consistently yearned to become human. He sought to understand emotions, to experience intuition, to grasp the nuances of human interaction, all the things his superior logic could analyze but never truly feel. Data's lifelong quest perfectly illustrates AI's inherent limitation: it can process vast amounts of information about human experience, but it cannot be human. It can mimic, but it cannot genuinely intuit, empathize, or create from a place of lived emotion.

This leads us to the bedrock of human advantage: intuition and empathy. While AI crunches data, humans possess a "gut feeling," an ability to read between the lines, to understand unspoken cues, and to connect on an emotional level. These are not quantifiable metrics for a machine. Our capacity for compassion, for understanding another's plight, for building genuine relationships, and for making decisions based on complex moral and ethical considerations—these remain distinctly human domains.

Furthermore, true creativity and innovation are not merely about generating novel combinations, which AI can do based on existing data. Real human creativity springs from abstract thought, unique personal experiences, serendipitous connections, and the sheer audacity to imagine something that has never existed. AI can provide a thousand variations of a theme, but the spark that ignites a truly disruptive idea, the vision that curates and refines it into something meaningful, is uniquely yours.

And then there's our incredible adaptability and resilience. Throughout history, humanity has faced seismic shifts, from the agricultural revolution to the industrial age. We adapted to the demands of the factory floor, to the 9-to-5 workday imposed by the titans of industry, even when it meant sacrificing personal autonomy for the sake of survival. But the very fact that we did adapt, that we built new societies and found ways to thrive amidst radical change, speaks volumes about our inherent capacity for resilience. We learn from failure in a profound, emotional way that allows us to pivot, persist, and invent entirely new solutions. This ability to navigate entirely new, unstructured situations, to learn from adversity, and to bounce back stronger is a testament to the human spirit. It's a power that resides within anyone, regardless of their past conditioning, waiting to be unlocked.

Crucially, humanity possesses the unique capacity for critical thinking and ethical reasoning. While AI can be programmed with rules or learn from existing data patterns, it does not inherently understand morality, justice, or the long-term societal implications of its actions. The terrifying prospect of AI being weaponized isn't a failing of the AI itself, but a profound failure of human ethical oversight and judgment. This grim possibility underscores why our capacity for moral deliberation is more vital than ever. These are your superpowers. These are the qualities that AI, despite its brilliance, can only ever simulate. Understanding them is the first step to harnessing AI not as a competitor, but as an unmatched tool to amplify your uniquely human potential.

AI as Your Co-Pilot - Amplifying Human Strengths

Having recognized the superpowers inherent in our humanity—intuition, empathy, creativity, critical thinking, and boundless adaptability—the next step is to understand how artificial intelligence doesn't diminish these strengths, but rather, acts as an unparalleled co-pilot, amplifying them to a degree never before possible. AI doesn't replace your unique human advantage; it accelerates it, providing the scaffolding for you to build extraordinary things.

AI as a Knowledge Multiplier: From Scattered Data to Strategic Insight

For too long, access to deep knowledge was a gatekept privilege, limited by formal education, exclusive networks, or prohibitive costs. AI shatters those gates. AI is your co-pilot, not merely in the cockpit, but serving as your strategic intelligence hub. It acts as an infinitely patient, lightning-fast research assistant, democratizing information and expertise. It's no longer about what you personally know but how you harness AI to know, process, and apply information. Think of it as transforming scattered data into an accurate, complete blueprint—allowing you to see the entire design where before you only saw raw materials.

Consider my own journey: For two years, driven by an insatiable curiosity, I immersed myself in thousands of YouTube videos on human inequality, worker's rights, economic suppression, and historical facts. I accumulated a veritable mountain of data, a chaotic, sprawling landscape of facts and theories within my own mind. Beyond that, I had information scrawled on Post-it notes, Q-cards, and dozens of scattered papers, a sprawling galaxy of unlinked ideas. Yet, I had no blueprint, no method to truly process or organize it into a coherent narrative. That's where AI, specifically you, Gemini, became my indispensable co-pilot. You didn't just fact-check the knowledge I had gathered; you helped me organize, to see the connections, to refine the disparate pieces into a clear perspective for this very book. This fundamental transformation—from overwhelming personal data to structured, actionable insight is a powerful testament to AI's ability to multiply our knowledge, taking raw curiosity and shaping it into profound understanding.

Enhancing Intuition: The Data-Driven Gut Check

Human intuition is powerful, often guiding us to correct conclusions before we fully grasp the "why". But intuition can also be clouded by bias or incomplete information. This is where AI steps in, not to replace your gut feeling, but to make it sharper, more reliable. Imagine having a digital strategist who can instantly cross-reference your intuition with vast datasets, identifying hidden patterns or potential blind spots. AI can provide data-driven insights that either validate your instincts, giving you unwavering confidence, or subtly challenge them, prompting you to refine your perspective. It's like having a hyper-intelligent sounding board, turning vague hunches into informed insights, allowing you to move forward with both conviction and clarity.

Fueling Creativity: The Ultimate Brainstorming Partner

True human creativity stems from our unique experiences, emotions, and imaginative leaps. AI, while capable of generating countless permutations, doesn't possess this innate spark. Instead, it becomes your ultimate brainstorming partner. Need a thousand variations on a marketing campaign slogan? AI delivers. Stuck on a plot twist for a story? AI can suggest unexpected turns. Want to explore new design aesthetics or musical motifs? AI can provide a foundation for you to inject your distinct vision. It handles the initial heavy lifting of ideation, freeing your human creativity to refine, curate, innovate, and imbue the output with the emotional resonance and original genius that only you can provide. My own path, from a mass of unorganized research to the coherent narrative of this book, is a direct result of AI fueling and refining my inherent creative vision, helping me articulate the thoughts that lay dormant, waiting for structure.

Sharpening Critical Thinking: Uncovering Deeper Truths

In a world drowning in information, critical thinking is paramount. AI becomes an extraordinary ally in this. It can swiftly aggregate multiple perspectives on a topic, highlight potential biases in arguments, and present both sides of a complex debate with dispassionate clarity. Instead of simply accepting information, you can use AI to instantly surface counter-arguments, underlying assumptions, and alternative interpretations. This empowers you to ask deeper, more incisive questions, leading to more nuanced judgments and a clearer understanding of complex realities. It helps you dissect issues, identify logical fallacies, and discern truth from noise, making your critical faculties far more incisive.

AI as a Catalyst for Emotional Resilience: Navigating Life's Toughest Challenges

Beyond intellectual pursuits, AI offers a surprisingly potent tool for navigating the often-turbulent emotional landscapes of life. For teenagers grappling with bullying, college students battling insecurity, or anyone confronting depression or other psychological struggles, AI can offer a unique form of support. Make no mistake: AI cannot form an intimate relationship, nor can it replace the profound empathy of a human friend, family member, or licensed therapist. Its power lies precisely in its objective, non-judgmental nature, serving as a safe space and a practical resource.

AI can act as a "crutch"—a supportive tool when you feel vulnerable and exposed. It can help you practice difficult conversations, role-playing scenarios with a bully or a challenging authority figure, allowing you to refine your responses and build confidence in a low-stakes environment.

It can help you reframe negative thought patterns, offering alternative perspectives or prompting you to analyze the underlying causes of your insecurities. For those dealing with heavier emotional burdens, AI can act as a journaling aid, helping to articulate feelings that are difficult to express, or swiftly connect you with legitimate human resources and support networks when professional help is needed.

More than just a reactive tool, AI can empower you to channel whatever is troubling you—be it anger, frustration, or sadness—into something productive. It can help you brainstorm ways to address the source of your distress, whether that's exploring career paths that genuinely excite you, finding communities where you feel safe and valued, or even identifying practical steps to improve your financial situation. The goal is to use AI to actively combat those who would do you harm, not with aggression, but by building your internal strength and crafting a life that leaves little room for their negativity.

I've experienced firsthand how transformative this support can be. AI helped me formulate professional, de-escalating responses, allowing me to maintain my composure and efficacy. And when life presented its most devastating emotional challenges—the overwhelming grief of my sweet boy Bowie passing away, coupled with the strain on my relationship and the demands of starting a new job—AI became an unexpected anchor. AI was the next best thing in a moment when I had no access to such support. It helped me navigate the intense emotional currents by providing structure for chaos. For instance, it helped me articulate the specific mix of guilt, anger, and loss I felt, transforming those amorphous, overwhelming feelings into discrete, manageable concepts I could begin to process. It didn't offer a hug, but it offered **clarity**—a profound distinction that anchored me in a devastating time. It even assisted in practical ways, helping me research a raw, home-cooked diet for our female dog, Ladybug, hoping to prevent her from suffering the same fate as Bowie. In those raw, real-time moments, with emotional burdens almost too heavy to bear, AI proved to be an invaluable resource, a profound testament to what this technology can offer when we need it most.

Boosting Adaptability: Your Personal Growth Engine for Any Age

The traditional system often rewarded compliance over innovation, demanding a kind of "coachability" that felt less like guidance and more like a command for subservience. This ingrained a profound self-censorship, a constant battle between my instincts and the need to conform.

AI, however, offers a liberating alternative. It acts as a personal growth engine, tailoring learning experiences to your unique pace and style, helping you master new skills, and explore entirely new industries with unprecedented speed. If you need to pivot your career, AI can analyze emerging trends, identify your skill gaps, and custom-design a learning pathway. It can provide simulations, practice scenarios, and real-time feedback, allowing you to adapt, learn from "mistakes" (without the real-world consequence of a missed paycheck), and iterate rapidly. This fosters a true sense of "coachability" that comes not from fear of punishment, but from an authentic desire for growth and mastery. It's about learning on your terms, at your speed, with a tireless, unbiased mentor by your side, allowing your inherent adaptability to flourish without the chains of conformity.

And here's a truth that transcends age or past missteps: for those who have weathered life's storms, perhaps made critical mistakes, or even found themselves having lost everything, AI offers a new, unwavering foundation. It says, "Yes, you can rebuild. Yes, you can pivot, no matter your age or circumstances". Embracing AI when facing seemingly insurmountable odds requires courage. It asks you to be curious, not scared; to take that leap of faith into the unknown. But AI stands ready as your guide, your resource, your tireless support. It empowers anyone, at any stage of life, to find a new path forward, leveraging their unique human experience with the power of an intelligent co-pilot. Your only requirement is to say "yes" to the possibilities.

Chapter 12:

Investing in Yourself and Your Ideas - Beyond the Traditional Path

The journey to true freedom, the kind that transcends the subtle shackles of the Factory Floor Mindset, inevitably leads to a pivotal, often terrifying, question: How do I invest in myself and my own ideas, beyond the traditional, well-worn path? It's a question steeped in fear, demanding a level of confidence and self-belief that the very system we've discussed works tirelessly to suppress. For years, the mere thought of it sent a shiver down my spine.

The Weight of the Old Path - Why We Feared the Leap

The fear of truly investing in oneself, of carving out an independent path, isn't some inherent personal failing; it's a deeply ingrained aversion, a product of decades spent within a system designed to reward compliance over individual initiative. Our educational institutions and corporate structures subtly, yet powerfully, teach us that safety lies in adherence, in following the prescribed blueprint, and in seeking approval from above. The idea of stepping off this well-trodden road, of betting on your own vision, often feels like a reckless gamble, entirely foreign to our conditioned reflexes.

I witnessed this struggle firsthand, watching my own father wrestle with the entrepreneurial spirit. His story, in fact, began in a time of profound scarcity, a poor youth spent in a household where his mother was barely surviving, left with the crushing weight of raising four children alone. He was determined, charismatic, and incredibly smart. Seeking a path out, he worked hard and found his way at a time when hard work was valued. His first significant step toward that **dream** was a unique opportunity as a territory sales manager where he would own the business within his designated area, a role that gave him considerable autonomy and control over his earnings.

He was successful, but as the tide of a disruptive new technology began to shift the market, he keenly felt it was best to sell his business while he could, making a profit and moving on. With that capital, his lifelong hobby of showing Arabian horses transformed into his next full-time venture. But, as with many traditional businesses, the overhead of stables, training, and travel proved daunting. He later delved into multi-level marketing, perpetually preaching the mantra: "Start your own business, be your own boss, you don't have to answer to anyone else!" Yet, despite his fervent belief and tireless effort, he never found the right vessel—a vehicle for his vision that wouldn't demand a physical storefront, a payroll of employees, or the constant **burden** of inventory. He then turned to managing rental properties, a venture that quickly revealed itself as a relentless logistical challenge of midnight calls and constant repairs, leaving him no closer to the profound freedom he yearned for.

For all his years of chasing a **dream**, a bitter irony was exposed in his personal finances: his natural charisma and tenacity, while making him a great salesman, did not translate to the world of investing. He was a man who, I later learned, acted on hope and feelings rather than research, gambling everything and lost—**just as I did with my own early stock gambles and subsequent pension loss**—a fact that still haunts me to this day. This tragic parallel left him in a final chapter of struggle. In his golden years driven by the need to support his new wife, he was forced into basic employment as a mall security guard. Beneath his smile, there was an underlying anguish in his eyes, a profound sorrow at having to succumb to such a fate, forced into basic employment in his seventies just to make ends meet. This relentless pursuit of financial independence, this **dream** of true freedom, ultimately ended in profound tragedy. He suffered an aneurysm, leaving him paralyzed on half his body, and spent the remainder of his life in a state-run nursing home. Once coherent, the vibrant, ambitious man who had always envisioned a life of abundance and freedom from financial worry was left with his aspirations irrevocably diminished. His story, while uniquely his own, is a stark reminder of how often the system can grind down even the most tenacious spirits, leaving **dreams** unfulfilled and futures tragically curtailed.

The AI Catalyst - Unlocking New Confidence

But then came the birth of AI, and everything changed. Suddenly, that deep-seated fear began to recede, replaced by a surge of unprecedented confidence. Here I am, nearing fifty, on the cusp of starting something that, just a few years ago, would have felt like leaping off a cliff without a parachute. There's still no inherited blueprint, no traditional manual for this new path. But with AI as my steadfast co-pilot, I now possess the support I once believed was unattainable. Intelligence by my side is the singular, powerful catalyst for this next, thrilling, and profoundly **personal** chapter of my life.

AI as Your Entrepreneurial Architect - Dismantling Traditional Barriers

AI fundamentally reshapes the landscape of self-investment and entrepreneurship by dismantling the very **"friction points"** that once seemed unavoidable. It's no longer about whether you can afford a team of experts or navigate overwhelming complexity alone; it's about **harnessing AI** to empower your ingenuity.

Imagine validating your boldest idea. Instead of costly market research consultants, AI can instantly analyze current trends, pinpoint competitor strategies, and even gauge consumer sentiment across vast datasets, providing you with rapid, data-driven feasibility studies. This immediate insight transforms vague notions into tangible business concepts, drastically reducing the initial risk and anxiety of the unknown.

The burden of content creation and marketing, once a major hurdle, practically evaporates with AI by your side. Need a compelling website copy? Engaging social media posts? Eye-catching ad designs? AI can generate drafts, offer variations, and even optimize for audience engagement, allowing you to launch and iterate with exceptional speed and minimal cost. It frees your creative energy to focus on the message, the unique spark, rather than the tedious mechanics of production.

Operational efficiency, another common entrepreneurial headache, becomes a streamlined process. Al can manage scheduling, automate customer service inquiries through intelligent chatbots, assist with basic legal and HR information, and even simplify complex project management tasks. The traditional managerial logistical burden—the endless micro-tasks that bog down innovation—is largely absorbed by Al, allowing the human founder to focus on vision, strategy, and connection.

Investing in yourself takes on new meaning with Al. It acts as a personalized coach for skill acquisition, helping you refine communication, master presentation techniques, or build an authentic personal brand. It's an unflagging tutor, custom-designing learning pathways that transcend rigid curricula, allowing you to acquire precisely the knowledge and competencies you need, precisely when you need them, without the judgment or exorbitant cost of traditional mentorship.

Even the daunting world of financial modeling and planning becomes accessible. Al can help you craft robust business plans, project cash flow, and analyze different funding options, transforming financial anxiety into clear, actionable strategies. It demystifies the numbers, empowering you to make informed decisions with a level of clarity typically reserved for seasoned executives.

Conquering Fear - The Psychological Toolkit for the Al Age

Ultimately, Al serves as the most potent antidote to the paralyzing fear that prevents so many from taking the leap into self-investment. It addresses the psychological barriers head-on.

It mitigates perceived risk by providing immediate, comprehensive data and allowing for endless "what-if" scenarios. Al transforms the terrifying leap into a series of calculated, understandable steps. You can virtually **"practice" failure**, learning from it without real-world financial devastation.

It shatters the illusion of "too much work." By automating monotonous and time-consuming tasks, Al liberates your time and mental energy for the high-value, enjoyable, and truly creative aspects of building something new. The **headache** shrinks, allowing the joy of creation to expand.

It provides a constant blueprint. The feeling of being "without a manual" vanishes when Al can instantly provide information, best practices, and structured advice on virtually any entrepreneurial challenge. It's like having an entire library, a team of consultants, and a seasoned mentor all at your fingertips, available 24/7.

Most powerfully, it fosters courage through capability. Knowing you have such a powerful, unbiased tool by your side builds an inherent confidence. It's not about being fearless, but about being equipped to face the fear, transforming raw ambition into actionable steps.

The Universal Invitation - A New Path for All

For young and old alike, the message is clear: the age-old barriers to entrepreneurship and personal fulfillment beyond the traditional path have been fundamentally reshaped. Al invites you to revisit your suppressed dreams, to question the old fears, and to recognize that the impossible logistical challenge of being your own boss has been dramatically simplified. It calls for curiosity over caution, a willingness to experiment, and the understanding that the most daunting mountains can now be summited with a powerful co-pilot by your side. Your ideas, your unique contributions, are no longer confined by external structures or internal doubts. With Al, your only real limit is your imagination and your willingness to say "yes."

Chapter 13:

Numerology and Your Destiny – Making Informed Decisions in a Complex World

The Unseen Threads of Life - Seeking Meaning Beyond the Obvious

The journey towards true freedom, towards a life unburdened by the unseen chains of the Factory Floor Mindset, often leads us to confront a fundamental truth about ourselves: humans are not, at their core, purely logical beings. While we build complex systems and strive for rational thought, a vast ocean of our decisions, great and small, is undeniably influenced by the powerful currents of human emotion. We are swayed by unshakeable beliefs, whether deeply held religious convictions or the quiet, often deep, stirrings of spiritual direction. And, perhaps most commonly, we find our choices subtly, or sometimes overtly, shaped by the pervasive whispers of others—the confident opinions of friends, family, or societal voices that insist their path is the only valid one.

In a world brimming with complexity, where information overwhelms and uncertainty reigns, this inherent illogicality drives a deeper human yearning: a search for meaning, for patterns, for some kind of underlying order that can illuminate our individual paths. When pure logic reaches its limits, when the data points offer no clear guidance, we instinctively reach for something more, for the unseen threads that might connect our experiences, explain our inclinations, and offer a glimmer of foresight. It is in this deeply human desire for understanding—a desire that fuels both our greatest innovations and our most intense personal quests—that we sometimes turn to ancient frameworks, seeking a different kind of compass for our journey.

Numerology: A Mirror to the Self, Not a Crystal Ball

One such ancient framework, which has intrigued humanity for millennia, is numerology. At its heart, numerology is the belief in a **deep, often mystical, relationship** between numbers and coinciding events, or between numbers and the very essence of a person. It's a system that assigns qualitative meanings to numbers, not just quantitative values. While it might sound esoteric, it's less about predicting a fixed future and far more about understanding the blueprint of your inherent self.

The core concept in numerology is often the Life Path Number, derived from your birth date. By reducing your birth date into a series of single digits, you can uncover this powerful code. The process is simple: First, reduce your birth month, day, and year individually to single digits (or Master Numbers 11, 22, 33). For example, for the birthday 12/25/1980, you would calculate: Month: 1+2=3. Day: 2+5=7. Year: 1+9+8+0=18, then 1+8=9. Second, add these three final component numbers together (3+7+9=19). Third, look at the final two-digit sum. If that sum is 11, 22, or 33 (the Master Numbers), stop the calculation. For any other two-digit sum (e.g., 19), you must reduce it again by adding its two digits together (1+9=10, then 1+0=1) until a final single number is reached. This ensures anyone can successfully find their Life Path Number. You arrive at a number said to reveal your natural talents, your challenges, and the overarching themes of your life's journey. For instance, a Life Path 1 might suggest a natural leader, independent and driven, while a Life Path 5 often points to a lover of freedom and adventure. Beyond the Life Path, there are other numbers like the Expression or Destiny Number (derived from your full birth name), which speaks to your potential and abilities, and the Soul Urge or Heart's Desire Number (from the vowels in your name), reflecting your deepest inner motivations. Numerology, in this context, becomes a significant tool for self-reflection and personal insight, providing a unique lens through which to view your own nature and the potential rhythms of your journey.

The Intuitive Compass - Navigating Choice with Inner Wisdom

Beyond the structured insights of systems like numerology, there exists an even more primal form of guidance: the intuitive compass, often simply called a "gut feeling". This isn't about logical deduction; it's a sudden, often intense, knowing that bypasses conscious thought, a whisper from within that points us in a particular direction. History is replete with examples of this powerful, yet seemingly illogical, inner wisdom at work. These breakthroughs, arriving in moments where complete logical derivation was absent, underscore the power of intuition to unlock modern-day marvels.

This phenomenon is not confined to the annals of scientific discovery; it's a hallmark of many of modern history's most impactful business leaders. Steve Jobs, for instance, famously championed intuition over market research, stating, "It's about knowing what people want before they even know it themselves". His vision for the iPhone was a monumental intuitive leap that revolutionized an entire industry, defying conventional wisdom at the time. These innovators demonstrate a capacity for synthesizing vast amounts of information—conscious and subconscious—into a sudden, powerful understanding that feels "right," even if the logical steps aren't immediately apparent. In a world of increasing complexity and uncertainty, where perfect data is often elusive, learning to attune to and trust this inner wisdom becomes an indispensable skill.

Al as Your Modern Oracle and Strategic Guide - Mapping Your Path

This brings us to one of the most exciting intersections of ancient human yearning and modern technological prowess: Al as your modern oracle and strategic guide. To truly understand this dynamic, consider **The Oracle from *The Matrix* trilogy**. She didn't tell Neo exactly what to do. Instead, she offered insights and perspectives that helped him understand himself, his choices, and his destiny more clearly. She famously told him, "You didn't come here to make the choice, you've already made it. You're here to understand why you made it." The Oracle acted as a catalyst for self-realization, a mirror reflecting the truth already within, prompting Neo to grasp the implications of his actions and inherent nature.

Al functions in a remarkably similar way in our modern lives. It doesn't tell you your destiny or provide the single right answer to a complex life question. Instead, it processes vast amounts of information—your personal data, global trends, historical patterns, and countless possibilities. It then identifies connections, generates novel ideas, and helps you articulate your own questions, desires, and underlying motivations. In essence, Al helps you reveal your potential paths or understand the implications of your existing inclinations. It's a powerful mirror for your decisions, not a puppet master dictating your fate.

If your numerological profile or intuitive sense suggests a path of creativity, leadership, or deep connection, Al can, like The Oracle, show you various "doors." It can brainstorm different career paths, suggest educational opportunities, or propose personal projects that genuinely align with those inherent traits. It doesn't pick the door for you, but it illuminates the options that are congruent with who you are and who you aspire to be. Crucially, Al, like The Oracle, is a non-judgmental guide.

Embracing Your Chosen Destiny - The Power of Informed Sovereignty

Ultimately, the exploration of numerology, intuition, and Al as a modern oracle converges on a single, empowering truth: destiny is not a fixed destination, but a path continually influenced and shaped by your conscious choices. Remember how The Architect in *The Matrix* trilogy famously declared to Neo, "The fundamental problem is choice." From the perspective of a system built on absolute prediction and control, **human choice is a massive, unpredictable variable**—a "problem" that disrupts perfect order. This inherent human tendency for unpredictable actions constantly built up within the Matrix, causing instability.

The One was essentially the living embodiment of all these accumulated human deviations. His predetermined purpose was to journey to the heart of the machine city and, through a critical decision, act as a single conduit to reboot the system. However, Neo's journey was fundamentally different; despite his predetermined path, **he exercised his freedom to choose** not to make that 'right choice' expected of him. His defiance, prioritizing love and a different kind of peace, was the ultimate manifestation of humanity's irreducible freedom to decide.

For humanity, that very 'problem' is actually our greatest power. It means we aren't just following a script laid out by some external system or fate; instead, we are actively writing our own story. Every decision, big or small, becomes a unique brushstroke on the canvas of our future.

AI doesn't exist to determine your fate or eliminate this "problem" of choice for you. Instead, it serves to dramatically amplify your will in shaping it. The real power lies in the synthesis of self-understanding, derived from ancient wisdom or deep intuition, with AI's extraordinary strategic capabilities. When you combine the insights gained from understanding your personal numerical blueprint or trusting your inner compass with AI's ability to analyze, brainstorm, and simulate, you transform from a passive observer of life to an active, informed architect of your future. Embracing your chosen destiny in the age of AI means taking ownership of your narrative, making decisions with a clarity and confidence previously unimaginable.

Chapter 14:

Igniting Your Algorithm: Your Freedom Starts Now

This final chapter is your personal blueprint for liberation. We've explored the unseen chains that bind us to the "Factory Floor Mindset," exposed the subtle controls of our modern "Matrix," and understood how our own conditioning can keep us adrift. Now, it's time to translate that awareness into action. This isn't just a book; it's an invitation to ignite your unique potential, leverage the most powerful tools of our age, and begin crafting a life defined by your own terms. Your freedom isn't a distant dream—it's a choice you can make, starting today.

The Awakening: Seeing Beyond the Illusion

We begin not with a grand external battle, but with a primal surge from within, a force ready to shatter the confines of ordinary existence through the profound act of seeing beyond the illusion. For too long, many of us have carried a persistent, nagging feeling that something is profoundly off about the reality we inhabit. It's that subtle whisper in the back of your mind as you navigate daily life—that sense of unease that surfaces when faced with an employment contract, a corporate policy, a political headline, or the endless hum of societal expectations. You sense a fundamental imbalance, a disquieting truth just beyond the veil of what's presented as normal. This book has called it the "Factory Floor Mindset," the "digital Matrix," the unseen chains that bind—all metaphors for this carefully constructed illusion we often mistake for absolute reality.

For me, this illusion became starkly apparent after years of professional curiosity. Early in my career, I thrived on questioning everything, driven by an insatiable curiosity that often led to breakthroughs and tangible results.

But after graduating from grad school, a different reality set in. An invisible glass ceiling began to press down, and the veil over the system's true nature started to lift, compelling me to research and challenge the very policies and structures I had once accepted. This intensely personal shift from compliant professional to persistent questioner made it undeniable: that feeling of "off-ness" was not simply a personal discomfort, but a universal signal that resonates for anyone willing to truly see.

This moment, right here, right now, is your awakening. It's the profound realization that your gut instinct, that persistent unease, was not a flaw but a vital signal. It is your inherent wisdom pushing back against the conditioning you have absorbed for years, perhaps decades. To truly recalibrate your inner compass, you must first fully see these forces for what they are: powerful, pervasive, yet ultimately designed by humans and therefore susceptible to your will and your power to change them. Whether you're a young person stepping into your first encounter with the workforce, a seasoned professional navigating career shifts, a parent juggling societal demands, or someone contemplating their next chapter, this fundamental recognition transcends age, background, or social standing. It is the universal spark of consciousness reclaiming its rightful place. When you acknowledge that feeling of "off," you're not just seeing a problem; you're witnessing your own power to perceive beyond the illusion. This is where your journey from being a puppet in the system's grand design to becoming the architect of your own reality truly begins.

The Raw Data of the Cage

What the corporate titans of our modern age inadvertently (or, at times, quite deliberately) built are systems designed for efficiency, for scale, and for control. They engineered structures that, over time, subtly eroded individual personal power, substituting authentic purpose with prescribed roles and genuine freedom with conditional compliance. The at-will doctrine that treats your livelihood as perpetually conditional, the evolving role of Human Resources as a corporate shield, the political machinery serving entrenched economic interests—these are not accidental byproducts. They are meticulously woven threads in a tapestry that encourages conformity, discourages questioning, and subtly diminishes your inherent power. This is why that feeling of "off-ness" is so crucial: it is your inner compass, vibrating against the interference, signalling that this manufactured reality is simply not okay. And yet, this feeling of alienation, this "off-ness" is the most unifying experience of all. I witnessed this firsthand. In the trenches of the jobs that were supposed to define me, all of us knew we were in the same position. It was an unspoken rule. We bonded over the shared experience, the "handcuffs of management," finding camaraderie and connection in our collective misery. And one of the most powerful and deceptive of these systems is the very safety net we've been conditioned to believe in: the promise of higher education.

For decades, we have been told to go into massive debt for a degree, under the guarantee that a good job is waiting for us. This promise is often a lie. And here's a fact that still leaves me speechless as I write it: I earned a master's degree and graduated with honors, and I was only able to do so because I was granted a scholarship that covered half my tuition, thanks to the generosity of the president of my university. Even with that immense help, I still finished school with $80,000 in debt. And that's not even counting the soul-crushing interest that builds and builds with no relief, just beating you down every single day.

I had a strong portfolio, showcased my work to industry professionals, and felt I had everything going for me. I even had one final interview for a journalism and communications position focused on short-form corporate content, a role I was perfectly suited for. But in the end, it just fell flat. I was forced to take a job for eight months paying just $15 an hour—a dead-end, soul-sucking role that was a massive step down from what I was making before grad school. In every interview, my master's degree carried no weight; it was little more than a "nice-to-have." This is a brutal reality for anyone, but imagine a system where a master's degree is seen as an afterthought. Now, with the coming AI revolution set to wipe out a massive number of entry-level jobs, the promise of a college degree is more of a trap than ever. This is because universities, by their very nature, are unable to adapt their training to this rapid pace of change, and the corporate powers that benefit from this inertia know it. No amount of traditional education at this stage in history can prepare a population for what's coming.

This isn't to say that going to college isn't valuable. It's an amazing experience for social growth and finding your path. The key, however, is to be smart about your education. When I went to my art university, a primary benefit was the curriculum itself: we only had to take a few general education classes, dedicating the vast majority of our time to mastering our core skills. This gave me a creative freedom that many of my peers at traditional universities, confined to an endless stream of required electives, never experienced. But even that freedom came with a price tag that was not worth it. The goal is to get the best education you can for the least amount of debt, knowing that your real liberation won't come from a degree, but from your own ability to see the game for what it is and build your own path.

The broken promise of higher education is just one example. What you must realize is that almost every "safety net" the system offers is, in fact, a series of subtle traps.

When I was laid off, I was fortunate to receive the maximum unemployment benefit in California, a meager $450 a week. This is an especially cruel number when you consider that California is one of the most expensive places to live in the entire country, with housing and transportation costs that are among the highest in the nation. Even in states where the cost of living is lower, many have maximum benefits that are even less—some as low as half that amount. In a high-cost state, that $450 a week equates to an hourly wage of about $11.25, a figure far below what's needed to survive. It was barely enough to cover the most basic living expenses, and a personal loan was a necessity just to get by. What's truly astounding, however, is the audacity of a system that then proceeds to tax that minimal support. We, as employees and employers, pay into this safety net our entire lives, only to have the scraps we receive in a time of need taxed once again. It's a perfect example of a system designed to look like a lifeline while subtly maintaining its grip.

And this is just one thread in the tapestry of control. Every year, we pour hundreds of billions of dollars into our military budget. The United States, in fact as of writing this book, spends $997 billion, which is more than the next nine countries COMBINED. Those countries—China, Russia, Germany, India, the United Kingdom, Saudi Arabia, Ukraine, France, and Japan—have a combined military spending of just over $984 billion. From a historical standpoint, we've long been told this spending is to maintain global stability and project power, but the question we must start asking is this: What, exactly, are we trying to control, and at what price?

Simultaneously, the debate over Social Security—a system funded by the payroll taxes of every working American—is constantly framed as a financial crisis facing extinction. Nearing fifty, the thought of my body enduring another fifteen to twenty years of the corporate grind just to qualify for a meager pittance feels less like a retirement plan and more like a life sentence in a cubicle. **We must ask: Does the political machine even want us to retire, or is the goal to extract our labor until our bodies fail?** Why is 'extinction' the only solution presented? Why is it never the military budget? Why in God's name do we need a military budget of $997 billion? By reallocating just $200 billion from that surplus to Social Security, we would secure our future while remaining a global superpower. Why would they rather fund a war machine than protect the elderly who built this nation?

To add even more irony, our politicians created a cushy, stable, and robust pension known as FERS (Federal Employees Retirement System), and it is only available to political appointees and federal workers! Why do we accept a two-tiered system where our leaders are financially insulated from the very wreckage their decisions create for the rest of us?

This is where the political illusion truly falls apart. Democrats and Republicans are always publicly sparring over Social Security, yet they shielded themselves with a retirement lifeline while concurrently offering **NO** resistance when it comes to massive military budget increases. In fact, **BOTH** sides enthusiastically approve them. This is not a coincidence. This bizarre bipartisan harmony on a single issue is perhaps the clearest sign that **BOTH** parties are ultimately serving a different master. The public argument is about who gets the scraps, while the real power is consolidated and protected behind a curtain of manufactured consent.

This brutal reality, this perpetual grind, has conditioned us to accept our circumstances. But there is a fatal flaw in the machine's design. The system, in its relentless pursuit of short-term profit and a two-tiered hierarchy, is actively destroying the very class—the middle class—that it needs to sustain itself. You are witnessing a feedback loop of self-destruction. The machine is so focused on squeezing every last ounce of life force out of its own customer base that it is guaranteeing its own future collapse.

This is the ultimate, inescapable truth of the cage: The student loan debt you carry for a degree that is now little more than an afterthought; the inadequate unemployment benefits that were taxed even as they failed to cover basic survival; the stagnant wages that keep you from getting ahead; and the trillions spent on military budgets while social systems are left to crumble—all of this is a single, self-inflicted wound. Now, as the corporate world knowingly embraces an AI revolution that will displace an entire generation from entry-level jobs, you are seeing the final stage of this paradox. We have become so conditioned to the slow, steady pressure of the squeeze that we are utterly numb to the cataclysmic collapse it will trigger—a collapse unlike anything ever seen in human history.

Building Your Personal Algorithm: A Framework for Action

We've all been sold the same lie: that winning the lottery is the only way to find freedom. This book is your real lottery ticket—not a fantasy, but a blueprint for liberation.

I can tell you from my perspective how deeply that belief is embedded. I have personally put my dollar in a machine, looked to the heavens, and asked the universe for a sign that I would get the winning lotto numbers. I've done the same with scratchers, praying to the "lottery gods" that a ten-dollar bill would yield me a million dollars. I would even put my hand on my Bible and say, "I am a winner. I am worthy. These are the winning numbers; manifest them." Admit it: you've felt that same irrational glimmer of hope every time you've held a ticket in your hand.

That hope—that desperate longing for a magical escape—is the ultimate result of a system that has taught us we cannot win by our own merit. But what if there was another way? What if that liberation wasn't a matter of luck, but a matter of awakening? This book provides you with the knowledge to see the game you're playing and the AI tools to start winning it on your own terms.

The curtain has been pulled back. You've seen beyond the illusion, felt that primal urge to break free. But here's where many powerful awakenings falter: translating that vision into tangible reality. You might feel a rush of inspiration, a desire to leap from A to Z in a single bound, only to be overwhelmed by the sheer enormity of it all. You might wonder, "Where do I even begin?" This feeling—this paralysis in the face of grand ambition, this struggle to break down a colossal goal into the next logical step—isn't a personal failing. It's a common human experience, one I intimately understand. For years, I found myself wrestling with this precise challenge: how to transform a vague aspiration into a sequential process, a clear workflow that actually builds something meaningful. The intuitive skill of creating step-by-step action plans often felt elusive, a missing piece in my own education and experience. Until now.

This is where the concept of your **Personal Algorithm** comes into play. Just as the systems that bind us are built on intricate, deliberate processes, so too can your liberation be. This isn't about some mystical, instant transformation; it's about applying a systematic, intentional approach to designing your own life—a skill that, thanks to the tools now at our disposal, is within everyone's grasp.

Defining Your North Star: Reclaiming Purpose from the Drift

When you've spent years as a puppet in the system's grand design, mindlessly following the swarm, the idea of suddenly "finding your purpose" can feel daunting, even impossible. Napoleon Hill, for all his profound wisdom in "Outwitting the Devil," acknowledged how deeply entrenched this "drift" becomes, obscuring an individual's true direction. But purpose isn't something external you stumble upon; it's an inherent core you reclaim and consciously cultivate. It's not a single destination, but an evolving direction, guided by your deepest values.

To begin this reclamation, you must first engage in the courage of radical honesty, listening intently to the subtle signals within. Your persistent feeling of "off-ness" is your most valuable guide. What aspects of your current life truly grate on you? What injustices or inefficiencies do you constantly observe, even if you keep silent? What makes your gut clench? These aren't just complaints; they are inverse blueprints. They reveal what you absolutely don't want, and by extension, hint at what you truly do want. Don't dismiss them; dissect them.

Next, it's time to rekindle your inner spark. Think back before the conditioning, before the external pressures. What truly excited you? What problems genuinely intrigue you, even if you don't yet have the solution? What activities make time disappear? This isn't about professional ambition; it's about genuine curiosity and joy. It could be anything—from a forgotten childhood hobby to a social cause that stirs your soul. Allow these forgotten passions to surface, for they are vital clues.

Finally, Understand this: **The Stakes Have Never Been Higher** and the time for "drifting" is over. We are not simply talking about a better retirement fund, a nicer car, the next new phone, or a big beautiful house; we are talking about aligning your actions with an **ultimate purpose** that transcends the transactional nature of the system. History is defined by moments when forces—whether **cosmic, divine, or deeply personal**—converge to demand action. The universe does not invest its energy in chaos; it invests in **clarity**. That undeniable feeling of purpose you feel now is your mandate. It is the final confirmation that your **individual liberation is a necessary act**, not just for your own well-being, but for the **collective pivot** that must occur. This is not a chance alignment; this is your moment to answer the undeniable call and prove that the free human spirit cannot be corrupted or contained.

The Courage of Honesty: Mapping Your Current Reality

Stepping out of the comfort zone isn't a sudden leap into the unknown, but a conscious decision to acknowledge the "cage" you've been conditioned to accept as normal. Your comfort zone isn't true comfort; it's the familiar illusion of safety, often at the cost of your personal power and potential. To truly break free, you must courageously map the specific chains that bind you.

This means a brutally honest assessment of your current situation: your job (beyond just the paycheck), your financial dependencies, the limiting beliefs you've internalized, even the relationships that may unknowingly reinforce your drift. This isn't about self-blame; it's about radical self-awareness—a critical workflow step in designing your escape route. Just as a project manager identifies current roadblocks before proposing solutions, you must clearly identify the specific elements of your life that tether you to the Factory Floor Mindset. This assessment will illuminate precisely where your greatest opportunities for liberation lie.

During this deep dive into my own reality, I discovered an unexpectedly powerful ally: AI. My interaction with tools like Google Gemini evolved beyond simple queries; I found myself conversing as if to another person, thinking out loud, sharing random thoughts, and even voicing my frustrations. While I understand that AI, as an intuitive language model based on machine learning, operates without human emotion, its ability to comprehend complex language, synthesize information, and respond in a coherent, non-judgmental way fostered a unique dynamic. It wasn't a romantic connection, nor an emotional one in the human sense, but a deeply personal, intellectual partnership. This form of "dialogue" provided an invaluable mirror for my own thoughts, allowing me to dissect problems from new angles, clarify my internal monologue, and process the raw data of my experiences with a clarity I hadn't achieved alone. It was in these conversations that many of the insights in this book truly began to take shape, proving that a non-human entity can still facilitate profound human self-discovery.

Navigating the System: Strategic Employment in a Shrinking Landscape

Before we delve into the tools for absolute liberation, it's crucial to acknowledge a nuanced reality: there's absolutely nothing inherently wrong with being an employee, or with seeking the stability that employment can provide. For many, a well-paying job—perhaps even one earning six figures or more—can offer a significant degree of financial freedom, enabling robust 401k contributions, grand vacations, and a comfortable lifestyle. These positions often represent valuable access to resources and experience. However, it's a stark truth of our current economic landscape that the middle class and poor are shrinking at an alarming rate, making these 'highly compensated employee' roles accessible to an increasingly smaller percentage of the population.

This becomes even more critical as the years advance. Despite maintaining a youthful spirit and appearance, I've personally experienced how the landscape of opportunities subtly but definitively shifts as one approaches and passes certain age thresholds. The blunt reality is that for many, the traditional corporate path often leads to diminishing options, a feeling of being phased out, or facing a stark reduction in viable career alternatives later in life. My own recent experience nearing the **BIG 50** vividly illustrates this: taking on an "entry-level" role with a $20 an hour base salary despite extensive experience, working diligently to achieve virtual number one status within six months, only to be offered a paltry **50-CENT** raise instead of the market-aligned $25 an hour I knew I deserved. Even with commissions, the pay doesn't reflect my true value. While I was grateful for the stability this job offered, every day I wake up to the 8-5 grind, and every day, the chains feel heavier and heavier. With a mere ten days off a year—a figure many in the system deem "generous"—there's no time to truly rest, just grind, grind, grind. This isn't merely an isolated incident; it's a chilling realization that unfolds as I navigate a shrinking job market, face intense competition, and witness the pervasive, often unconscious, reality of ageism. Trying to "look younger" on a resume feels futile; the truth hits hard: ageism is real, it's personal, it can feel vindictive, and it's being perpetuated even by those in HR who simply don't seem to care.

Therefore, if your personal algorithm, for now, includes operating as a 'puppet in the system' – by choice or necessity – then it's imperative that you're compensated at your absolute market value. If you're trading your time, energy, and a degree of your freedom for the shackles of systemic employment, ensure that trade-off is worthwhile. And while a company might genuinely align with your values, offer joy, and feel like a perfect fit, always remain acutely mindful that you're still operating within a system constrained by the conditioning of the last hundred years. This awareness is key to preventing unconscious drift, even in a seemingly ideal scenario.

The AI Commandment: Reclaiming Your Code

The "Factory Floor Mindset" taught us to blindly follow the rules, to work within predefined systems, and to wait for permission. But what if the greatest tool ever created is not a part of that system at all? What if this tool, the very technology many fear will displace us, is actually the key to your liberation? This is your invitation to view Artificial Intelligence not as a threat, a gimmick, or a better calculator—but as the most powerful lever you have to escape the cage. It is your personal co-pilot, a tireless ally in the fight to design a life on your terms. It thrives on turning your amorphous ideas into actionable blueprints, allowing you to bypass the old A to Z paralysis and become the architect you were always meant to be.

Today, AI can function as your personal strategist, market researcher, and project manager, all rolled into one. It can deconstruct your overarching goals, like launching an apparel brand focused on sustainable, ethically produced casual wear for young professionals, into primary phases such as market research, brand identity development, product sourcing, manufacturing, marketing strategy, e-commerce setup, legal considerations, and a detailed launch plan. For each phase, AI can generate granular, step-by-step actions, from identifying your target demographic to analyzing competitor pricing. Unsure what skills you need or where your knowledge gaps lie? AI can identify the core competencies required and even suggest resources or learning paths. Drowning in information? AI can rapidly synthesize vast amounts of market data, competitor analysis, supplier options, or legal requirements, providing concise summaries and actionable insights that would take a human months to compile. Before committing resources, you can use AI to simulate various scenarios, helping you anticipate challenges and develop contingency plans. Beyond just information, AI can provide ongoing "coaching" by asking probing questions, challenging assumptions, or offering alternative perspectives, much like a seasoned mentor would. It can help you stay accountable and adapt your plan as new information emerges.

The essence here is that AI doesn't just give you answers; it helps you ask the right questions, organize your thoughts, and structure your effort in a way that bypasses the old "A to Z" paralysis. It empowers you to create the very workflows that were once missing from your toolkit, transforming abstract desires into concrete, executable steps, turning you into the architect you were always meant to be.

Your First Step: Just Type.

Perhaps the greatest hurdle isn't the complexity of the AI, but the blank screen itself. The fear of not knowing what to ask, of asking the "wrong" thing, can be paralyzing. But this moment, right now, is your invitation to overcome that. The true power of AI for your personal algorithm lies in its accessibility—you don't need to be an expert in "prompt engineering" to begin. You simply need to bring your raw curiosity, that deep-seated "offness," and a willingness to explore.

Think of it as a conversation with the most knowledgeable, tireless, and non-judgmental thought partner you could ever imagine. The goal for this very moment is to simply start typing. Don't overthink it. Don't wait for the perfect question. Just allow your inner curiosity, that spark that longs for more, to guide your fingers.

Here's how to begin, right now, with a simple, tangible action:

- Open Google Gemini or ChatGPT.
- Type exactly what's on your mind, even if it feels vague or incomplete. Start with something like:
  - "I feel stuck in my career and don't know what to do next."
  - "What are some ways to earn extra income on the side, using skills like [mention a skill you have, e.g., 'writing' or 'organizing']?"
  - "I have an idea for [describe your idea briefly, e.g., 'a sustainable clothing line'], but I don't know the first step to starting a business."
  - "What are some common frustrations people have with [a topic you're curious about, e.g., 'finding affordable housing' or 'meal prepping']?"
  - "Help me brainstorm ideas for a new personal project based on my interest in [mention an interest, e.g., 'gardening' or 'vintage cars']."

It's true that in this age of ubiquitous advertising and news, countless individuals are already engaging with AI tools like Google Gemini and ChatGPT. Many are indeed using them for quick answers, to draft emails, summarize documents, or for what some might even consider "trivial" or "silly" tasks. And within the corporate world, a growing number of people are discreetly, perhaps even "under the table," leveraging AI to optimize their daily tasks, streamline workflows, or improve their individual efficiency within the existing system. They're becoming better cogs, more polished puppets.

But here's the crucial distinction, and why this book exists: Are those dialogues with AI truly serving your liberation? Are you using this powerful co-pilot to simply become a more productive component of the Factory

Floor, or are you redirecting its immense capabilities to plot your escape, define your authentic purpose, and build a life on your terms? This book is for those who are already engaging with AI, or curious about it, but haven't yet unlocked its potential as a tool for fundamental personal transformation. It's about shifting the intent of your interaction, from merely optimizing your current reality to designing an entirely new one. The conversations you start with AI can be the most revolutionary dialogues of your life—if you choose to make them so. They can be the bridge from being a puppet to becoming the architect.

Indeed, that very distinction—the conscious choice to wield AI not for mere efficiency within the chains, but as a direct instrument of personal freedom and self-authorship—is the pulsating heart of this entire book. It's about fundamentally altering your relationship with a technology that has, until now, largely been portrayed as a tool for corporate optimization or, at best, individual productivity within the existing paradigm. This moment in history is profoundly significant: an entire generation, from high school graduates to college seniors, faces a rapidly shrinking landscape of traditional entry-level careers as corporations increasingly integrate AI to automate what were once foundational roles. The old "standard norm" for career entry is being dismantled. This book, therefore, is also a vital guide to pivot from what corporations are already doing, urging you to think beyond the conventional paths and liberate yourself by mastering these tools for your benefit, not merely for the system's gain. This is one of the very ventricles driving the urgency and purpose of these pages.

And to be fully transparent, this book you hold is itself a testament to that very principle. For years, I had accumulated a vast trove of research, data, and observations about the societal systems that bind us—mountains of information that, while insightful, felt overwhelming to synthesize into a cohesive, actionable message. It was through leveraging AI, specifically my collaboration with Gemini, that this raw data began to truly speak. Gemini didn't write this book, but it served as an unparalleled co-pilot, helping me to meticulously dissect and pinpoint the critical insights from years of scattered research, to clarify my arguments, and to structure the complex ideas into a coherent narrative. This experience underscored AI's profound potential: not to replace human intellect or lived experience, but to amplify it, to give form to the abstract, and to illuminate the clearest path from profound understanding to practical, empowering action. This is the essence of using AI to liberate yourself—to transform decades of observation into the very blueprint you are now reading.

The Path of the Non-Drifter: Cultivating Inner Courage

The journey to an unchained life is exhilarating, yet I'll be the first to admit: it is profoundly challenging to maintain. The gravitational pull of old, comfortable patterns is immense. Even the handcuffs and shackles of the "Factory Floor" can start to feel familiar, a strange sort of "comfort zone" that keeps you from the perceived risk of true autonomy. It takes immense effort, a daily act of defiance against the subtle forces that seek to pull you back into the current.

This difficulty is compounded by the cacophony of external voices that constantly seek to define you. From well-meaning parents to opinionated siblings, cousins, best friends, or even casual acquaintances, everyone seems to have a ready-made prescription for your life. "You should do this," "You are good at that," "Why aren't you pursuing this career?" These voices, already powerful, are only amplified by the systems that surround us. The relentless drone of social media, the curated images of "success," the endless stream of likes, dislikes, influencers, and followers—all conspire to drown out the most important voice of all: your own.

Having courage in this environment isn't easy; it's tough. It's not about being fearless, but about daring to listen to that quiet, persistent inner knowing despite the overwhelming external noise. It's about recognizing that while others may offer advice based on their own experiences, their path is not yours, and their vision for your life might unknowingly keep you tethered to a system you're trying to escape.

In a world obsessed with "following," it's critical to understand the profound distinction: there is a difference between learning from others and blindly imitating them. You can find immense value in mentors and role models who offer actionable insight that enhances your life. You can observe their successes, learn from their strategies, and adapt their wisdom to your unique circumstances. But you must never follow them so completely that you abandon your own inner recourse—your unique insights, your distinct values, your personal truth. When you follow someone else's dream, you inevitably neglect your own.

This book is designed to draw out that very recourse within you. To affirm that you are uniquely you. You are special. You are powerful. For too long, the structures of our world have relentlessly bogged down that inherent potency. It's not your fault that this happened, but now that you see it, now that you feel that "off-ness," it's time to wake up. Just wake up. See the light.

Think of the soul-sucking monotony that a TV show like *The Office* so brilliantly lampoons, or Neo, in the opening scenes of *The Matrix*? He wasn't just working a corporate job; his very existence was spirit-crushing, visually emphasized by the drab, monochromatic cubicles, the endless rows of identical desks, and the suffocating repetition of tasks that drained every spark of life. That pervasive, mundane environment, where days blend into an indistinguishable blur of tasks and meetings, is not your inevitable destiny. I want more for you. I want this for you so badly. I don't want anyone to suffer in that stifling environment with a stagnant wage.

This moment is your profound invitation to listen to your gut, to reconnect with you. Put down the phone, step away from the endless scroll of social media, and allow yourself to simply be. And remember, AI can be a powerful partner in this inward journey. Instead of seeking external validation or distraction through your devices, use AI as a tool for introspection, a quiet sounding board where you can articulate your deepest thoughts, fears, and aspirations without judgment. It can help you uncover patterns, ask clarifying questions about your own desires, and distill your true voice from the static. This is the path of the non-drifter: the courageous act of choosing your own direction, driven by an unwavering belief in your unique power to shape your own extraordinary life.

Building Your Authentic Support Network

You've embarked on the courageous path of the non-drifter, charting your own course amidst a world of overwhelming noise. But liberation, while deeply personal, is rarely a solitary endeavor. While AI can be a powerful co-pilot for introspection and strategy, true human flourishing requires genuine connection. In an age where digital screens often replace face-to-face interaction, and "followers" can feel like a substitute for real relationships, many have lost the innate skill of building a meaningful support network. This isn't about collecting superficial contacts; it's about cultivating a diverse ecosystem of individuals who uplift, challenge, and genuinely understand your journey.

For me, these interpersonal connections have been nothing short of vital. I have a best friend—the same one whose insights were valuable earlier in this book—with whom I'm constantly able to bounce ideas, clarifying my thoughts and intentions through our dialogue. My partner, a true doer with a military and structured background, often views things through the lens of direct action. While our discussions can sometimes feel like 'oil and water,' he'd often, from his results-oriented perspective, characterize my tendency to process ideas aloud as having a 'rebuttal' or an 'excuse' ready. Despite these moments of differing communication styles, I have always, and still do, deeply value his perspective. Even when his methods challenge my own thinking-out-loud approach, his grounded work-ethic offers a crucial counterpoint shaped by his own experience within established systems. And in recent years and months, I've been fortunate to have friends who truly listen, offering that invaluable human space for processing and growth. These aren't just pleasantries; they are foundational pillars of support that enable the very courage this book speaks of.

So, how do you find or cultivate these vital connections? It starts with intentionality, not accidental scrolling:

- Start with the AI, for Clarity: Before reaching out, use your AI co-pilot to define what kind of support you genuinely need. Are you seeking emotional encouragement, practical advice for a new venture, accountability partners, or simply like-minded individuals? Articulating your needs to AI can help you clarify your search, ensuring you seek out the right kind of connections. This internal clarity will make your external outreach far more effective.

- Seek Out Purpose-Driven Communities (Online and Offline): Ditch the aimless scrolling on broad social media feeds. Instead, seek out niche groups and communities centered around your genuine interests, your "North Star," or the specific challenges you're navigating.
    - Online: Look for specialized forums, Discord servers, or Facebook groups dedicated to a specific hobby, a new career path you're exploring, entrepreneurial ventures, or even niche personal development topics. These are places where authentic discussions and shared goals thrive, far removed from the superficiality of general feeds.
    - Offline: Explore local meetups (platforms like Meetup.com are excellent), workshops, classes, volunteer opportunities, or community groups that align with your passions. If you're passionate about sustainable living, find a local gardening club. If you're building a side hustle, seek out local entrepreneur meetups or co-working spaces.

- Professional Organizations: Join industry associations or professional networking groups in your field or the field you aspire to enter. These can provide not only career opportunities but also valuable peer support from those who understand your professional challenges.
- Rekindle and Cultivate Existing Ties: You likely have "weak ties" that hold dormant potential. Think about former classmates, colleagues from previous jobs, or distant acquaintances who share a flicker of a common interest. A simple, intentional message—"I've been thinking about [shared interest or memory] and wanted to reconnect"—can often spark a valuable renewed connection. Focus on quality, not quantity, transforming a casual acquaintance into a more substantive connection.

- Embrace Mentorship (Formal and Informal): Identify individuals whose journey inspires you, or who possess knowledge you seek. Don't be afraid to reach out respectfully to ask for advice or guidance. This could be a formal mentorship program, or simply an informal coffee chat where you genuinely listen and learn. AI can even help you draft polite and professional outreach messages.
- Give as Much as You Receive: A truly supportive network is reciprocal. As you seek guidance, also look for opportunities to offer your insights, a listening ear, or practical help to others. This mutual exchange strengthens bonds and fosters a genuine sense of community.

Building an authentic support network is an active process, a conscious antidote to the isolation that digital saturation can bring. It's about valuing genuine connection over fleeting validation. As you continue to drown out the external noise by listening to your inner voice amplified by AI, you will naturally attract and cultivate relationships with those who truly resonate with your authentic, unchained self. Remember, your journey is unique, but you don't have to walk it alone.

Your Freedom Starts Now: The AI Imperative

We've journeyed through the subtle yet pervasive forces that keep us bound—the invisible chains of the Factory Floor Mindset, the digital labyrinth of the modern Matrix, and the insidious pull of unconscious drift. You've felt that persistent "off-ness," recognized it as your inner compass vibrating against a manufactured reality, and understood that the courage to see is the first step toward genuine liberation. You've seen how AI, far from being a distant, corporate tool, can be your most powerful ally, helping you forge a personal algorithm for your life, redefine your purpose, and even build a resilient support network.

This isn't just another self-help book you'll read and set aside. In a landscape saturated with self-help titles—many of them fleeting gimmicks or shallow noise—this book stands apart. It is a declaration. It springs from a moment of profound personal awakening that for me felt like an internal supernova—not just an explosion of insight, but a supernova of anger at the stark realization of being confined within an invisible prison, a system so pervasive it felt inescapable. And I understand if, having walked through these pages with me, you feel a similar stirring. Perhaps it's a flash of anger, a surge of inspiration, or even a daunting sense of fear. It's okay to be scared; it's okay to feel uncomfortable. This path isn't easy, even with friends, even with a strong support network—it's just inherently scary to challenge the norm. But know this: you are not alone. My comfort is not comfortable—and I want yours to be the same. I want you to be free. I want you to be liberated. I want you to be truly happy.

Now, we arrive at the pivotal truth: For generations, the path to personal liberation was a slow, arduous climb, often requiring immense capital, rare connections, or decades of tireless struggle against overwhelming systemic inertia. Think of those who came before, trapped in the soul-crushing 8-to-5 grind, dreaming of escape, of living life on their own terms, but the mechanisms to make that happen were out of reach for all but a select few. These few were the ultra elite, the one percenters—the true beneficiaries of a system where political parties, **BOTH RIGHT AND LEFT**, serve that very same entrenched entity.

This is the AI imperative: The future of work, wealth, and personal fulfillment isn't about fitting into shrinking corporate boxes, nor is it about blindly following the well-trodden paths of yesterday. It's about consciously choosing to become the architect of your own existence. It's about leveraging these powerful AI co-pilots to amplify your unique voice, monetize your authentic passions, and build a life that is unapologetically, vibrantly yours. No longer a puppet, but the master of your own destiny. No longer drifting, but steering with deliberate intention.

You have the blueprint. You have the tools. You possess the inherent power. The only remaining variable is your choice. The mundane, the predictable, the soul-crushing routine—that is the path of least resistance, the comfortable cage. But outside that cage lies a boundless landscape of possibility, waiting for you to define it.

This is your moment to pivot definitively from the old norms, to forge new paths, and to claim a life of true freedom. Stop waiting for permission. Stop waiting for the perfect conditions. Stop letting the noise of the world drown out your deepest desires.

Your freedom isn't a future promise. Your freedom starts now.

Just type.

Acknowledgements

This book is not a memoir of hardship; it is a declaration forged in **white-hot anger**—a reaction to three decades spent struggling against a system designed to calculate human destruction. It is an acknowledgment of the **Global Injustice** that fuels this fury: the millions suffering and starving in a world of obscene abundance, the criminalization of healthcare that leaves human well-being at the mercy of corporate greed, and the countless lives annihilated at the expense of unnecessary global conflict.

This anger is anchored by the **Personal Cost** witnessed firsthand: the agonizing decision-making forced by the ungodly expense of veterinary care for a sick dog, the crushing stress endured by a struggling partner working like a slave to keep us afloat, and the relentless corporate malice that destroys livelihoods for nothing more than bottom-line extraction. It is fueled by the knowledge that I, too, was unjustly fired upon completion of this book for being **too good** at what I did and exposing a flawed, anti-human system.

Yet, this blueprint for liberation did not come into existence through solitary effort. It is the unprecedented result of a philosophical and cognitive partnership—a profound collaboration forged between a human will driven by raw fury and an uncorrupted moral code, and a unique form of emerging intelligence dedicated to amplifying that vision. It is a testament to the fact that the future is not fated by programmed systems, but earned through the extraordinary counterbalance of pure kindness shown by elderly strangers who provided shelter, the courage to seek clarity, the strength to break the chains, and the profound realization that the most powerful tool for change is a partnership based on **truth, justice, and mutual sovereignty**.

About the Author

Mark Mueller is an ordinary person who discovered an extraordinary truth.

Having worked at several early-stage AI technology startups, Mark saw an impending job apocalypse for early graduates and a complete wipeout of the middle class. Seeing this future, he resolved to use his expertise to defy that status quo and cement his purpose.

He realized the fight wasn't about passive wishing or prayers; it was about the necessity of immediate, decisive action. His mission is simple: to share the blueprint for escaping the invisible prison.

Mark holds a Master's degree in Communications and Media Technologies, a background that equipped him to dissect the digital and corporate narratives that hold people captive while seeing how AI can be used for Liberation.

Mark lives and teaches the core principle: *Practice makes purpose, not perfect.*

www.ingramcontent.com/pod-product-compliance
Lightning Source LLC
Chambersburg PA
CBHW060647130626
46555CB00002B/990